The Japanese Culture of Mourning Whales

Mayumi Itoh

The Japanese Culture of Mourning Whales

Whale Graves and Memorial Monuments in Japan

palgrave
macmillan

Mayumi Itoh
Princeton, New Jersey
USA

ISBN 978-981-10-6670-2 ISBN 978-981-10-6671-9 (eBook)
https://doi.org/10.1007/978-981-10-6671-9

Library of Congress Control Number: 2017956576

Cover design by Fatima Jamadar

Printed on acid-free paper

This Palgrave Macmillan imprint is published by Springer Nature
The registered company is Springer Nature Singapore Pte Ltd.
The registered company address is: 152 Beach Road, #21-01/04 Gateway East, Singapore 189721, Singapore

Dedicated to the souls of whales

Notes on the Text

All translations were made by the author in the form of paraphrases (not as literal translations) in order for the translations to make sense in English. For instance, the word "*kuyō-tō*" literally means a "memorial tower," but this book gives it different translations depending on the nature of the monument. When the monument was erected in a temple and was made in the style of a stupa (a memorial stone in Buddhism), the word is translated as "memorial stupa." In turn, when a monument was created in a park or at the shore, it is translated as "cenotaph" or "memorial tower" (for a taller one).

Regarding spelling in Japanese, the Hepburn style is primarily used, with macrons, but macrons are not used for words known in English without macrons, as in Kyoto and Tokyo. Another exception is that "n" is not converted to "m" for words where it precedes "b, m, and n"; for example, Tenpō era, instead of Tempō era (names for Japanese newspapers, such as *Asahi Shimbun*, are given 'as is' because they are their official English names). Also, names of Japanese persons are given with the surname first, except for those who use the reversed order in English. Honorific prefixes, such as doctor and mister, are not used in the text.

As with reading archival documents, deciphering the inscriptions on whale gravestones and monuments presented a challenge (many of the engravings have weathered away and become illegible). In fact, the more one reads documents on whale graves, the more contradictions and discrepancies one finds. The study is complicated by the fact that the date of the inscriptions was recorded in the old calendar system, in which the imperial era name, as well as the lunar calendar, was used. The imperial era frequently changed during medieval times, and there are as many as thirty-six imperial eras for the Edo period (1603–1868) alone. Many of the names are similar and are easy to confuse with one another. Also, the lunar calendar usually lags a month (sometimes two months) behind the Gregorian calendar. For this book, the date was translated into the Gregorian calendar wherever possible.

In addition, the old age counting system (*kazoe*) used in Japan confuses the dates of birth of Buddhist monks or the dates of memorial services given for whales by Buddhist temples. For instance, the third-year anniversary memorial service is actually held on the second-year anniversary day of the death. For this book, sources were cross-examined and the *kazoe* date was adjusted to the Gregorian calendar wherever possible. When the exact date could not be determined, this book provides two dates, as in 1654/1655.

Further, the recent nationwide drastic changes in names of places add much confusion to the research. Due to the "Great Mergers of the Heisei [Era]" that began in 1999 and peaked in 2005, a number of existing cities, towns, and villages were abolished and absorbed into newly created ones. As of April 13, 2016, there are a total of 1718 cities, towns, and villages in Japan—790 cities, 745 towns, and 183 villages. This total number is a 1514 decrease from March 31, 1999—a 120 increase for cities, a 1249 decrease for towns, and a 385 decrease for villages. The problem is that most of the newly created municipalities were named without much regard to the names of the previous ones with roots that reflected their local history.[1]

Consequently, the names for the new municipalities have lost the "colors of history" and traditional identities, and it is difficult to figure out the location of places by their new names. As documents on whale graves and monuments record the locations throughout Japan with the pre-Heisei merger names, matching the old names with the new unfamiliar ones presented a challenge.[2]

Citation numbers for sources of information are normally given at the end of each paragraph, instead of at the end of each sentence, in order to enhance the smooth reading of the text and also to limit the number of citations. Regarding online sources, the dates of actual access are given unless the sources give the posting dates.

Notes

1. "'Heisei no dai-gappei' tettei tsuiseki" (Full Investigation of "Great Mergers of Heisei [Era]"), April 13, 2016, http://www.8toch.net.gappei/
2. Ibid.

Map: Whale Graves and Related Monuments in Japan

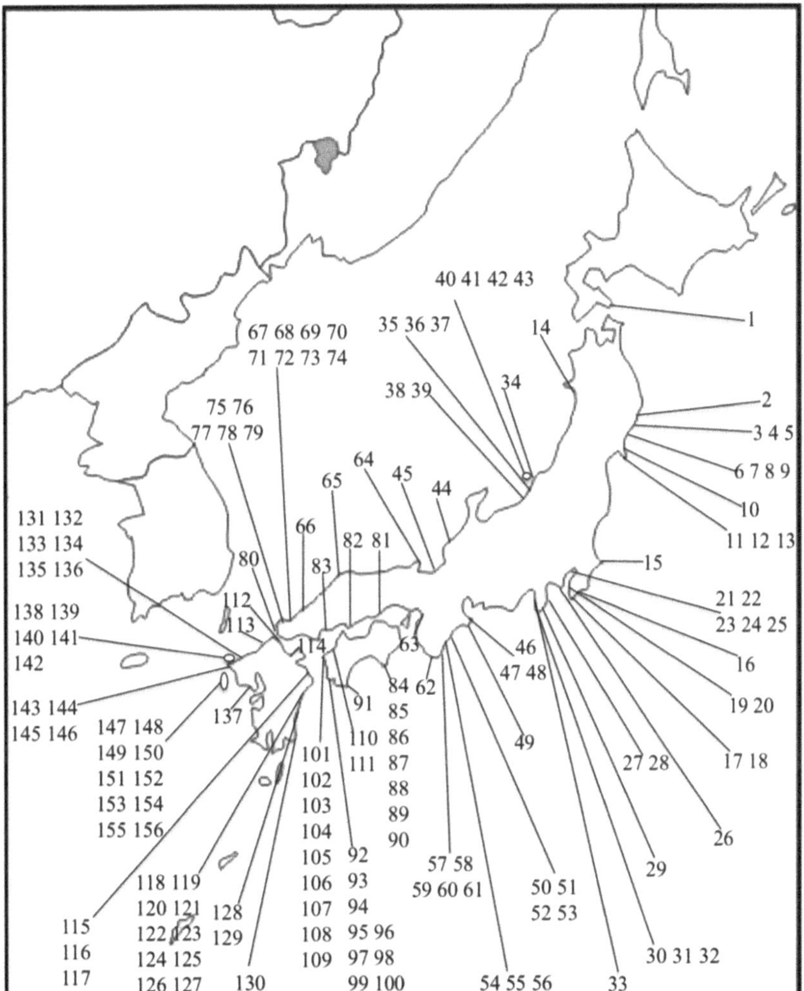

Note: The numbers correspond to those in Table A.1
Source: Constructed by author from blank map of East Asia, courtesy of Daniel Dalet, http://d-maps.com/carte.php?num_car=77&lang=en

Acknowledgments

Among the many people who encouraged me in writing this manuscript, I would like to thank Sara Crowley-Vigneau, Rachel Crawford, and Connie Li for editorial work; Fujii Fuminori, Furuyama Keikō, Hayakawa Yoshikatsu, Kawamura Shigeyoshi, Ken Kawata, Kumi Katō, Mori Tetsuo, Nakazono Shigeo, Ochiai Kanji, Seiji Ohsumi, Sakurai Hayato, Sasaki Katsutaka, Suda Keiko, Kyōko Suda, Usui Tsutomu, and Yasunaga Hiroshi for valuable information; and Daniel Dalet for the loan of a map. I also thank Tsuneo Akaha, Kent Calder, Toshiko Calder, Steve Clemons, Gerald Curtis, Joshua Fogel, Sheldon Garon, Ronald Hrebenar, Donald Keene, Ellis Krauss, David Leheny, Mike Mochizuki, T. J. Pempel, Stephen Roddy, Gilbert Rozman, Richard Samuels, Vicki Wong, Donald Zagoria, and Quansheng Zhao, as well as Gregory Rewoldt and Megumi Itoh, for continuous encouragement and inspirations.

Contents

Abbreviations

ICJ	International Court of Justice
ICRW	International Convention for the Regulation of Whaling
IWC	International Whaling Commission
JNR	Japanese National Railways
JR	Japan Railways
SCAP-GHQ	General Headquarters of the Supreme Commander of the Allied Powers
STCW	Small-Type Coastal Whaling
UNESCO	United Nations Educational, Scientific and Cultural Organization

List of Photographs

List of Tables

CHAPTER 1

Introduction

In nature one form of life must always prey upon another. However, human consciousness holds an awareness of, and sympathy for, the will of other beings to live. An ethical human strives to escape from this contradiction so far as possible.
—Reverence for Life, Albert Schweitzer (1923)

On March 31, 2014, the International Court of Justice (ICJ) ruled that Japanese whaling operations in the Antarctic Ocean were not for "scientific research" but were de facto commercial whaling. Japan thus lost the case that the Australian government had filed in May 2010. The Japanese government responded immediately by stating that it would abide by the ruling. This is not news. Earlier in 1982, the International Whaling Commission (IWC), based in Cambridge, England, which was established under the International Convention for the Regulation of Whaling (ICRW) in 1946, adopted a regulation to implement a moratorium on commercial whaling beginning in the 1985–1986 season. However, along with other whaling countries such as Norway and Russia, the Japanese government, which had joined the IWC in 1951, objected to its decision.[1]

The IWC has no power to enforce its decisions on its members, and any member nation can opt out of any specific IWC regulation by submitting a formal objection to it. Canada even withdrew from the IWC entirely in 1982. Japan in the end succumbed to international pressure (of the United States specifically) and withdrew its objection and stopped commercial whaling completely in the Antarctic in 1986. Nevertheless, Japan soon began whaling

M. Itoh, *The Japanese Culture of Mourning Whales*,
https://doi.org/10.1007/978-981-10-6671-9_1

operations in the Antarctic in 1987 under the "scientific research" provision of the ICRW (which also allows whaling under the "aboriginal-subsistence" provision). The "scientific research" provision allows governments to issue special permits to their citizens to engage in whaling for purposes of scientific research. Japan also began small-type "scientific research" whaling in the Japanese coast and off northwestern Pacific Ocean in 1994.[2]

Thus, unlike other whaling countries, such as Norway, Iceland, and Russia, that had opted out of the moratorium on commercial whaling and have engaged in commercial whaling on their own, Japan has engaged in "scientific research" whaling through the management of a government-supported non-profit organization, the Institute of Cetacean Research based in Tokyo. In turn, international environment protection groups, such as Greenpeace and the Sea Shepherd Conservation Society, condemned the Japanese government asserting that it was allowing commercial whaling in the name of the "scientific research." Their anti-Japanese whaling campaigns escalated to the extent that they attacked the Japanese whaling factory-ship Nisshin-maru in the Antarctic Ocean in 2006 and 2007. Then came the ICJ's ruling of 2014. These controversies are public knowledge.[3]

Japanese Classical Whaling

Little known is the fact that the Japanese have had a tradition of whaling since ancient times. The *Man'yō-shū*, the anthology of poems that was compiled from the late seventh century to the late eighth century, uses the word *isana-tori* ("*isana*" literally means "brave fish" and refers to whale, while "*tori*" means "to catch") in describing the ocean. This suggests that the Japanese had already engaged in a rudimentary form of whaling by the Nara period (710–794). However, the practice remained sporadic.[4]

Centuries later, with the invention of manual harpoons in the late sixteenth century, Japanese classical whaling began in earnest at the beginning of the seventeenth century during the Edo period (1603–1868). Unlike modern Western-style whaling, Japanese classical whaling only involved coastal whaling, catching whales that migrated off the Pacific Ocean or the Sea of Japan and using only rudimentary tools. Moreover, what is almost unknown to the non-Japanese speaking world is the fact that the Japanese conducted ceremonies after the deaths of whales and took care of their "afterlives."[5]

Japanese Culture of Mourning the Deaths of Whales

The memorial rites for whales began with creating whale graves by burying a part of their bones and erecting gravestones. Then, the Japanese conducted funeral services and mourned their deaths according to the Buddhist percept. Afterwards, they performed seasonal and annual memorial services for whales and continued to pray that their souls would rest in peace in heaven. They also erected memorial stupas (cenotaphs created at Buddhist temples and elsewhere) for whales and continued to pray that their souls would attain the enlightenment of Buddha. Surprisingly, these memorial rites have been observed not only in whaling communities but also in non-whaling communities where villagers caught whales that had been stranded ashore. Even more surprisingly, after the classical whaling had ended more than a hundred years ago, the Japanese have maintained this tradition to this day.[6]

Literature Review

Literature in English

There is a diverse literature in English on whaling in general and on Japanese whaling in particular. Some of the notable recent works on whaling include *The International Politics of Whaling* (1997) written by Peter J. Stoett; *Trying Leviathan: The Nineteenth-Century New York Court Case That Put the Whale on Trial and Challenged the Order of Nature* (2007) and *The Sounding of the Whale: Science and Cetaceans in the Twentieth Century* (2012), both by D. Graham Burnett; and *Unveiling the Whale: Discourse on Whales and Whaling* (2009) by Arne Kalland. Meanwhile, two books in English have been published on Japanese whaling: *Japanese Whaling: End of an Era* (1992) by Arne Kalland and Brian D. A. Moeran, and *Whaling in Japan: Power, Politics, and Diplomacy* (2009) by Jun Morikawa.[7]

There are also two book chapters on Japanese whaling: "Japanese Perceptions of Whales and Dolphins" by Arne Kalland in John Knight, ed., *Wildlife in Asia: Cultural Perspectives* (2004); and "The Ambivalence of Whaling: Conflicting Cultures in Identity Formation" by Jessamyn R. Abel in George M. Pflugfelder and Brett L. Water, eds., *JAPANimals: History and Culture in Japan's Animal Life* (2005).[8]

None of the entries above mention Japanese belief systems about whales or memorial rites for whales, with the exception of works by Arne Kalland.

In addition, there is a conference proceeding entitled *Small-Type Coastal Whaling in Japan: Report of an International Workshop* published by the Boreal Institute for Northern Studies, the University of Alberta (1988), which has a short chapter, "Whaling Beliefs and Japanese Worldview." Although this publication does not provide the author's name for each chapter, the twelve participants in this workshop include such prominent anthropologists and Japanologists as Tomoya Akimichi, Pamela J. Asquith, Harumi Befu (deceased), Theodore C. Bestor, Milton M. R. Freeman, Helen Hardacre, Arne Kalland, and Brian D. A. Moeran. The significance of this highly specialized study is unquestionable, but some of the content is now outdated. More recently, *Bones of Contention: Animals and Religion in Contemporary Japan* (2012) by Barbara R. Ambros touches on the subject.[9]

Accordingly, there is no book in English, to the knowledge of this author, that has devoted its entirety to the study of Japanese whaling culture in general, let alone the Japanese memorial rites for whales specifically. Regarding academic journal articles, only one article in English has examined Japanese memorial rites for whales for one specific locale, "Prayers for the Whales: Spirituality and Ethics of a Former Whaling Community—Intangible Cultural Heritage for Sustainability" (2007) by Kumi Kato.[10]

Literature in Japanese

In comparison, literature on Japanese whaling in general as well as on the Japanese whaling culture abounds. Regarding classical literature, *Saikai geigei-ki* (Record of Whaling in West Sea), compiled by Tanimura Yūzō in 1720, is considered the oldest solid study of whaling, chronicling classical whaling also in the Kinki and Inland Sea regions. This was followed by the publication of *Geishi-kō* (Manuscript on History of Whaling), compiled by Ōtsuki Seijun in 1808. This six-volume work is the oldest encyclopedia of whaling in Japan.[11]

In the latter twentieth century, notable publications came out, which include *Nihon hogeishi-wa* (History of Japanese Whaling, 1960) by Fukumoto Kazuo; *Saikai geigei-ki* (Record of Whaling in the West Sea, 1980) by Hirado-City Board of Education; *Kujira to hito no minzoku-shi* (Ethnography of Whales and People, 1994) by Tomoya Akimichi; *Kujira to hogei no bunka-shi* (Cultural History of Whales and Whaling, 1994) by Morita Katsuaki. This was followed by *Kujira-tori no keifu: Gaisetsu Nihon hogei-shi* (Genealogy of Whaling: Introduction to the History of Japanese Whaling, 2001) by Nakazono Shigeo; *Kujira to Nihonjin* (Whales and the Japanese, 2003) by

Ohsumi Seiji, then director general of the Institute of Cetacean Research; and *Kujira sono rekishi to bunka* (History and Culture of Whales, 2005) by Komatsu Masayuki, former official at Japan's Fisheries Agency and acting Japanese representative to the IWC; and *Kujira-tori e-monogatari* (Picture Tales of Whaling, 2009) by Nakazono Shigeo and Yasunaga Hiroshi.[12]

Further, Watanabe Hiroyuki added a sociological perspective to the subject in *Hogei-mondai no rekishi shakaigaku: Kingendai Nihon ni okeru kujira to ningen* (Historical Sociology of Whaling Problems: Whales and People in Modern Japan, 2006). In addition, there are a plethora of documents, such as the official history of local classical whaling, compiled by local governments respectively, where classical whaling bases had existed.[13]

Regarding the Japanese culture of mourning the deaths of whales specifically, Shindō Naosaku made a pioneering study in *Seto-naikai no kujira no kenkyū* (Study of Whales in the Inland Sea, 1968) and *Seto-naikai shūhen no kujira no kenkyū* (Study of Whales Surrounding the Inland Sea, 1970). Following in the footsteps of Shindō, a scholar at Tokyo University of Fisheries (currently Tokyo University of Marine Science and Technology), Yoshihara Tomokichi visited most of the fishing communities in Japan that are known to have preserved whale graves and monuments and updated the data collected by Shindō. Yoshihara's extensive field research culminated in "Kujira no haka" (Graves of Whales), which was compiled in *Nihon minzoku bunka shiryō shūsei* (Collection of Documents on Folklore Culture in Japan, 1997) edited by the renowned Japanese anthropologist Tanigawa Ken'ichi.[14]

In his painstaking and tireless study, Yoshihara identified about sixty whale graves and monuments in Japan. However, he omitted some of the significant ones because information was limited when he conducted his research. Actually, the total number in earlier times would have been more than 100, because many have decayed and disappeared.[15]

Scope of This Book

The purpose of this book is not to study modern Japanese whaling. Instead, this book examines the lesser-known aspects of Japanese whaling, specifically the cultural and religious aspects of Japanese classical whaling regarding how Japanese mourned the deaths of whales. Given the paucity of studies in English of Japanese classical whaling and its culture, this book also mentions other aspects of Japanese whaling culture, such as folklore legends about whales, as well as the history of Japanese classical whaling, and thereby fills the vacuum in studies in English of Japanese whaling. For this purpose, this book investigates various forms of memorial rites for whales, as manifested in whale graves, cenotaphs, and other monuments.

Unfortunately, many of the whale graves and monuments do not exist today because they have weathered away due to exposure to high waves and inclement weather; however, this book tries to be as exhaustive as possible (updating the existing data collected by Yoshihara Tomokichi and others) and studies all of the known cases for which records and relics of memorial rites for whales have been preserved. This author visited about a dozen fishing communities in Japan (most of which were located in remote regions with difficult access by public transportation) and interviewed officials at local whale museums and temple head priests. There are also monuments for other marine mammals, including dolphins (marine mammals were considered species of fish). This book includes these cases because information about them is rare.

It will be best to examine the subject matter in the eight regions of Japan from northeast to southwest—in the order of the Hokkaidō Region, the Tōhoku Region (northeastern Honshū, the largest island of Japan), the Kantō Region (central-eastern Honshū), the Chūbu Region (central Honshū), the Kansai-Kinki Region (central-western Honshū), the Chūgoku Region (western Honshū), the Shikoku Region (Shikoku Island), and the Kyūshū Region (Kyūshū Island). This study excludes Okinawa prefecture, because the region was an independent state—the Ryūkyū Kingdom—and few documents are available about whaling there. It had a suzerain-vassal relationship with Satsuma province and the Tokugawa shogunate government after Satsuma province invaded the kingdom in 1609, but remained a tributary state of China. Then, in 1872, the Meiji government incorporated the islands into Japan as Ryūkyū province at first and then as Okinawa prefecture in 1879.

Similar names for gravestones and for memorial rituals for whales are cited throughout Chaps. 3, 4, 5, 6, 7, 8 and 9, and this might give an impression of repetitiousness. However, references to similar names are inevitable, because this is the first comprehensive study in English of memorial rites for whales in Japan. The existence of similar monuments and services in different regions is actually a testament to the fact that fishing communities in various parts of Japan shared the same sentiments toward whales and mourned in similar ways the deaths of the whales they had caught.

Japanese whaling culture is remarkably rich, and many of its "relics" are designated as tangible and intangible cultural assets of the nation or local governments. For instance, there are several picture scrolls vividly depicting scenes of classical whaling. One of them, *Nankai tokugei zue* (Picture of Whaling in the South Sea), is preserved at the New Bedford Whaling Museum in Massachusetts, which was a major American whaling base, along with Nantucket.[16]

Regarding intangible cultural assets, there are whale dances and whale songs, as well as folklore legends about whales in many fishing communities. Among them, the culture of mourning the deaths of whales stands out because it directly concerns such profound matters as reverence for life and benevolence for life and death that many religious people, including Albert Schweitzer, have contemplated. It is an epitome of how the Japanese dealt with the contradiction of taking the lives of other animals for their own survival and how they tried to coexist with nature.

* * *

The sorrowful relation between the whalers and whales is poignantly expressed in the poem written by Kaneko Misuzu (April 1903–March 1930), who grew up in a former whaling community in Senzaki, Yamaguchi prefecture (see Chap. 7):

> "Kujira-hōe" (Memorial Service for Whales)
> A memorial service for whales, in the late spring,
> when the flying fish are caught in the sea.
> When the sound of the bell at a temple by the beach
> reverberates across the sea,
> When fishermen in the village hurry to the temple by the beach
> wearing formal kimono jackets,
> A lone whale calf hears the sound of the bell in the sea,
> and cries out loud, missing its dead father and mother.
> I wonder how far the sound of the bell reverberates across the sea.[17]

(In reality, a whale calf cannot survive without its mother.) What was it about the belief system of Japanese fishing villages that infused a sense of benevolence into the deaths of the whales they had caught? What was the culture that nurtured the pure mind of Kaneko Misuzu, who embraced genuine empathy for whales? This book is a "pilgrimage" in search of that spirit.

Notes

1. "Whaling in the Antarctic (Australia *v.* Japan: New Zealand intervening)," March 31, 2014, http://www.icj-cij.org/docket/files/148/18136.pdf
2. Ibid.; Isao Miyaoka, "State-Society Relations Under Foreign Pressure: Two Contrasting Cases in the Japanese Fisheries Policy," http://www.dijtokyo.org/doc/dij-jb_11-miyaoka.pdf, downloaded February 24, 2016. For details of *gaiatsu* (foreign pressure) politics, see Kent E. Calder, "Japanese Foreign Economic Policy Formation: Explaining the Reactive State," *World Politics*, Vol. 40, No. 4, July 1988, 517–541.
3. Miyaoka; Ian Hurd, "Almost Saving Whales: The Ambiguity of Success at the International Whaling Commission," March 29, 2012, http://www.ethicsandinternationalaffairs.org/2012/almost-saving-whales-the-ambiguity-of-success-at-the-international-whaling-commission-full-text/
4. Nakazono Shigeo and Yasunaga Hiroshi, *Kujira-tori e-monogatari* (Picture Tales of Whaling), Fukuoka: Gen-shobō, 2009, 8–20; Kumano Taiji-ura hogei-shi hensan-iinkai, ed., *Kumano no Taiji: Kujira ni idomu machi* (Taiji, Kumano: Town That Challenges Whales), Tokyo: Heibonsha, 1965, 4–16.
5. Ibid. (both).
6. Boreal Institute for Northern Studies, ed., *Small-Type Coastal Whaling i00n Japan: Report of an International Workshop*, Edmonton, Alberta: Boreal Institute for Northern Studies, the University of Alberta, 1988, 53–54.
7. Peter J. Stoett, *The International Politics of Whaling*, Vancouver: UBC Press, 1997; Arne Kalland, *Unveiling the Whale: Discourse on Whales and Whaling*, New York: Berghan Books, 2009; D. Graham Burnett, *Trying Leviathan: The Nineteenth-Century New York Court Case That Put the Whale on Trial and Challenged the Order of Nature*, Princeton, N.J.: Princeton University Press, 2007 and *The Sounding of the Whale: Science and Cetaceans in the Twentieth Century*, Chicago, IL: University of Chicago Press, 2012; Arne Kalland and Brian Moeran, *Japanese Whaling: End of an Era*, London: Curzon Press, 1992; Jun Morikawa, *Whaling in Japan: Power, Politics, and Diplomacy*, New York: Columbia University Press, 2009.
8. Arne Kalland, "Japanese Perceptions of Whales and Dolphins," in John Knight, ed., *Wildlife in Asia: Cultural Perspectives.* London: RoutledgeCurzon, 2004, 73–87; Jessamyn R. Abel, "The Ambivalence of Whaling: Conflicting Cultures in Identity Formation," in George M. Pflugfelder and Brett L. Water, eds., *JAPANimals: History and Culture in Japan's Animal Life*, Ann Arbor, MI: Center for Japanese Studies, University of Michigan, 2005, 315–340.
9. Boreal Institute for Northern Studies, ed., "Whaling Beliefs and Japanese Worldview," 52–65; Barbara R. Ambros, *Bones of Contention: Animals and Religion in Contemporary Japan*, Honolulu: University of Hawai'i Press, 2012, 57–62.

10. Kumi Kato, "Prayers for the Whales: Spirituality and Ethics of a Former Whaling Community—Intangible Cultural Heritage for Sustainability," *International Journal of Cultural Property* (2007), No. 14, 283–313.
11. "Kujira: Nihon no koshiki-hogei to zusetsu" (Whales: Japanese Classical Whaling and Illustrations), 2011, http://www.lib.u-tokyo.ac.jp/tenjikai/tenjikai2011/kujira.html
12. Fukumoto Kazuo, *Nihon hogeishi-wa* (History of Japanese Whaling), Tokyo: Hōsei daigaku-shuppankai, 1960; Hirado-shi kyōiku-iinkai, ed., *Saikai geigei-ki* (Record of Whaling in the West Sea), Hirado: Hirado-shi kyōiku-iinkai, 1980; Akimichi Tomoya, *Kujira to hito no minzoku-shi* (Ethnography of Whales and People), Tokyo: Tokyo daigaku-shuppankai, 1994; Morita Katsuaki, *Kujira to hogei no bunka-shi* (Cultural History of Whales and Whaling), Nagoya: Nagoya-daigaku-shuppankai, 1994; Nakazono Shigeo, *Kujira-tori no keifu: Gaisetsu Nihon hogei-shi* (Genealogy of Whaling: Introduction to the History of Japanese Whaling), Nagasaki: Nagasaki Shimbunsha, 2001; Ohsumi, Seiji, *Kujira to Nihonjin* (Whales and the Japanese), Tokyo: Iwanami-shinsho, 2003; Komatsu Masayuki, *Kujira sono rekishi to bunka* (History and Culture of Whales), Tokyo: Goma-shobō, 2005; Nakazono and Yasunaga.
13. Watanabe Hiroyuki, *Hogei-mondai no rekishi shakaigaku: Kingendai Nihon ni okeru kujira to ningen* (Historical Sociology of Whaling Problems: Whales and People in Modern Japan), Tokyo: Tōshindō, 2006.
14. Shindō Naosaku, *Seto-naikai no kujira no kenkyū* (*Study of Whales in the Inland Sea*). Kobe: Kobe-shi ishi-kyōdo-kumiai, 1968 and *Seto-naikai shūhen no kujira no kenkyū* (*Study of Whales Surrounding the Inland Sea*), Kobe: Ikuta-ku ishikai, 1970; Yoshihara Tomokichi, "Kujira no haka" (Graves of Whales), in Tanigawa Ken'ichi, ed., *Nihon minzoku bunka shiryō shūsei* (Collection of Documents on Japanese Folk Culture), Vol. 18, Tokyo: San'ichi-shobō, 1997, 409–478.
15. Yoshihara.
16. Taiji-chōritsu kujira no hakubutsukan, ed., *Kujira-bune: Katachi to ishō* (Whale Boats: Styles and Designs), Taiji: Taiji-chōritsu kujira no hakubutsukan, 2011, 3; "New Bedford Whaling Museum," https://www.whalingmuseum.org, accessed February 24, 2016.
17. Kaneko Misuzu, *Kaneko Misuzu zenshū* (Complete Work of Kaneko Misuzu). Vol. 3, Tokyo: JULA shuppan-kyoku, 1984, 221–222.

CHAPTER 2

Historical Background

The Japanese engaged in classical whaling (*koshiki hogei*) for more than three hundred years, from the late 1500s to the early 1900s. Japanese classical whaling differed in many ways from modern whaling. The former was coastal whaling (*kinkaki hogei*) and thus was limited to operations in coastal areas using rowboats and rudimentary tools such as manual harpoons and hand-knit fishing nets. In contrast, the latter involves large-type, pelagic whaling (open-sea, long-distance whaling, or *enyō hogei*) employing a fleet of factory ships and several catchers equipped with bomb lance harpoons and guns, which were invented by Svend Foyn in 1870—the "Norwegian method."[1]

In addition, other than classical whalers, Japanese fishermen in non-whaling communities took part in so-called "passive whaling" as they caught whales that had been separated from their pods and had drifted into local bays, or those that were stranded ashore after being chased by orcas (killer whales). Pods of whales passed through the Pacific Ocean and the Sea of Japan during their migration seasons. They migrated from the south to the north in the spring after female whales gave birth in warm waters ("*nobori kujira*," the ascending whales) and then migrated back to the south in the fall ("*kudari kujira*," the descending whales).[2]

Accordingly, Japanese classical whalers engaged in coastal whaling only during whales' migration season. They were seasonal rather than fulltime whalers, who worked as ordinary fishermen for most of the year. They normally lived at a subsistence level, because the terrain in fishing communities was not suited to produce rice, the Japanese staple. This was also true

M. Itoh, *The Japanese Culture of Mourning Whales*,
https://doi.org/10.1007/978-981-10-6671-9_2

for most of the professional members of classical whaling guilds, with the exception of a few families that had organized large whaling guilds. Although there is a saying, "The catch of a whale made seven bay villages prosper," the prosperity brought by the catch was only temporarily.[3]

While a few *amimoto* (owner/managers of boats and fishing equipment) enjoyed a wealthy lifestyle during good seasons, they soon fell into debt during bad seasons because the maintenance of whaleboats and equipment cost dearly. In turn, local provinces imposed high taxes on catches of whales, because they brought an enormous amount of revenue, so whale guild heads often borrowed money from private lenders (an early form of banks) and also sought for financial assistance from the province.[4]

Four Major Classical Whaling Bases

Toward the end of the sixteenth century, active large-type whaling using the manual harpoon method (*tsukitori-hō*) emerged. Some of the earliest classical whaling guilds were organized in Owari (the western part of the current Aichi prefecture) around 1570–1573, in Ogawa Island (Saga prefecture) in 1594, and then in Taiji (Wakayama prefecture) in 1606. In the seventeenth century, the manual harpoon method spread through wider regions of Japan, and four major classical whaling bases were established: the Kishū Classical Whaling Base in Taiji (in current Wakayama prefecture), the Chōshū Kitaura Classical Whaling Base in Kayoi on Ōmi Island, Nagato (Yamaguchi prefecture), the Tosa Classical Whaling Base in Muroto (Kōchi prefecture) on Shikoku Island, and the Saikai Classical Whaling Base that encompassed Yobuko (Saga prefecture) and Ikitsuki Island (Nagasaki prefecture) in Kyūshū.[5]

Then, in 1675, Wada Kakuemon in Taiji revolutionized Japanese classical whaling by inventing a net method (*amitori-hō*), in which a number of chaser boats surrounded a whale on three sides and chased it toward large nets. This was much more effective and less labor-intensive than the manual harpoon method. As Wada did not monopolize his invention, the net method spread rapidly throughout western Japan, so that the net method combined with the harpoon method became the dominant form of whaling in the Tosa and Saikai Classical Whaling Bases.[6]

Meanwhile, whalers in the Chōshū Kitaura Classical Whaling Base developed a simpler net-drive method, in which they drove a whale into a bay and then closed the bay by casting nets. The fishermen in Ine, Kyoto prefecture, adopted this net-drive method. Elsewhere, in Katsuyama, in Awa (southern part of Chiba prefecture), whalers continued to use the manual harpoon method and maintained its tradition of catching Baird's beaked whales since the early seventeenth century, because Baird's beaked whales tend to dive deep in the sea and the net method was unsuitable to catch them.[7]

Whales Caught by Japanese Classical Whalers

Japanese classical whalers identified six major species of whales to be hunted: blue whales, fin whales, North Pacific right whales, humpback whales, sei whales, and sperm whales. Because they were easier to catch, the North Pacific right and sperm whales were most frequently taken (also by American whalers). Unlike larger whales, such as blue whales and fin whales, these medium-sized whales do not sink after being killed. The English name for right whales derives from the fact that they were the "right" whales to hunt, because they were easy to catch. In turn, the Japanese name for right whales is "*semi-kujira*" (*lit.*, "whale with a beautiful back"), because they have a smooth back without a dorsal fin. The Japanese also caught gray whales and Baird's beaked whales.[8]

The introduction of the net method enabled the Japanese to catch humpback whales—which are fast swimmers and sink after death—and sei whales. Nevertheless, blue whales and fin whales were simply too big for classical whalers, and they could only watch a pod of blue whales or fin whales migrating off the coast of Japan. Therefore, Japanese fishermen only caught blue whales and fin whales that were stranded ashore.[9]

Thus, as the Inuit in North America did for centuries, the Japanese fishermen-whalers engaged in seasonal whaling equipped only with rudimentary tools. It was a daunting task for small Japanese men to confront a huge creature in the sea. This was especially the case for the *hazashi* (*lit.*, "blade thruster"), who risked his life each time by being the first crewman to jump onto the whale's back, naked in the freezing water, to make a cut in the whale's blowhole through which a fishing net was tied. Because of the sheer size and might of whales, Japanese fishermen-whalers held them in awe and admired them.[10]

Today, the Inuit are still permitted to catch whales under the "aboriginal-subsistence" provision of the International Convention for the Regulation of Whaling (ICRW). Accordingly, the Japanese government has requested that small-type coastal whaling (STCW) conducted in several Japanese fishing communities be recognized as "aboriginal-subsistence" whaling under the ICRW. However, this was rejected. Nevertheless, there are certain similarities between Japanese classical whaling and Inuit classical whaling. After hunting, both shared the catch of whales with the community as a whole. The Inuit not only consumed whale meat but also utilized other whale parts such as baleen (long strips made of keratin that hang from the mouths of toothless "baleen whales"), from which they made sleigh boards and drinking cups.[11]

Similarly, the Japanese considered whales "blessings from the gods of the sea" and utilized every part of the whales they caught. They ate whale meat and used baleen to make a variety of items, ranging from combs, Japanese chess (*shōgi*) pieces, parts of string musical instruments, to signature seals. They extracted whale oil not only from blubber (fatty parts under the skin), but also from the bones by grinding and boiling them. They then used whale oil as fuel for lanterns and as lubricant for tools, as well as insect repellant in rice fields and houses (they coated walls of rooms with whale oil). They even ground the bones further, after extracting whale oil from them, and used them as fertilizer. They did not waste any part.[12]

Japanese Culture of Mourning the Deaths of Whales

The Japanese also mourned the deaths of whales with the community as a whole. The fishermen especially felt deep guilt when they caught a calf, because they knew that the mother whale always swam back to rescue it, even if she had initially swum away (probably in order to distract fishermen's attention from her calf), and they could catch the mother whale eventually. The mother whale fought back fiercely, though. Although this was their way of life, the fishermen could not help but feel deep sorrow. Being profoundly religious and also superstitious (they feared they might be punished for killing whales), they conducted various forms of religious rites for whales in order to atone for the killing. They not only made graves for whales (by burying some of their bones) and erected gravestones and other monuments for whales, but also conducted funeral services and annual memorial services for them.[13]

Moreover, many fishing and whaling communities commissioned local Buddhist temples to accord each whale a *kaimyō* (a posthumous "afterlife" Buddhist name) at a considerable expense, treating whales as if they were individual human beings. As a point of comparison, today a temple head priest receives from his parishioner (the family of the deceased) a "donation" of at least two to five thousand US dollars for conferring an afterlife name on the deceased, depending on the rank given in the afterlife name. Then, the funeral caretaker ordered a memorial tablet (*ihai*), on which the afterlife name was inscribed. At some locales, a collective memorial tablet for whales, rather than for a single whale, was created. In either case, the memorial tablet represented the souls of whales and was placed on the altar of a temple.[14]

Then, the head priest recorded the posthumous names of whales in the *kako-chō* (*lit.*, "book of the past"; a death register), which recorded the deaths of his parishioners—including afterlife names, real names, dates of death, and ages at death. This register was compiled according to the month of the year so that the names of the deceased who died in the same month were listed on the same page. This helped the head priest to perform the monthly memorial service of the deceased (if a person died on the first day of a month, the first day of every following month is the "monthly memorial day" of the deceased), as well as annual memorial services. The register also served as a vital record for local residents in Japan.[15]

The afterlife name, the memorial tablet, and the death register constituted a "trinity of paraphernalia" for conducting funeral services and memorial services for human beings. It is astonishing that many fishing and whaling communities commissioned local temples to accord whales memorial rites identical to those for human beings. After conducting funeral services for whales, the whaling guild or the fishing community held memorial services for whales regularly—monthly memorial day services, annual memorial day services, and seasonal memorial day services on the Spring Equinox, the Bon Festival on the week of August 15, and the Autumnal Equinox. All these memorial rites cost a considerable amount of money (euphemistically called "donations" to temples) for fishermen who normally lived at subsistence level.[16]

Remarkably, they have continued the practice every year to this day, even after they stopped whaling more than a century ago. In this sense, the memorial rites for whales constitute one of the most salient types of religious observances in Japanese whaling culture.[17]

Why did whalers and fishermen mourn the deaths of whales so profoundly? Why was this tradition prevalent throughout Japan even in non-whaling communities? Where did this tradition come from?

Japanese Animism and Reverence for Life

The tradition of mourning the deaths of whales is fundamentally associated with ancient Japanese animism, a belief that everything—living and nonliving—has souls of their own, and therefore everything in nature, including all forms of fauna and flora, must be respected. It regarded nature as a sacred and holistic existence that encompassed everything in it, rather than as an object for human beings to conquer. Human beings are only a part of nature, not standing at the peak of its pyramid. Human beings are blessed to live in nature, accordingly, they should live in harmony with nature, not destroy it. This belief system, stressing coexistence with nature and interdependence among all living things in nature, constitutes one of the basic philosophies of ancient Japanese animism.[18]

Consequently, the Japanese treated with respect not only animals (including fish), but also plants and trees, and even inanimate objects such as stones and paper. With this cultural and religious background, the Japanese received whales as blessings from the gods of the sea and consumed them with genuine appreciation.[19]

Whales as the Gods of the Sea

Another belief system that affected Japanese whaling culture is ancient Japanese mythology. Historically, the Japanese worshipped many indigenous gods in Japanese mythology (similar to the Greek mythology) that became the foundation of Shintoism, and one of them is the Hiruko God. According to Japanese mythology—as recorded in *Kojiki* ("Records of Ancient Matters," the oldest record of Japanese history, compiled in 712) and *Nihon-shoki* ("Chronicles of Japan," the second oldest record of Japanese history, completed in 720)—Hiruko (*lit.*, "leech child") was the first (or the third according to *Nihon-shoki*) child of Izanagi and Izanami, who were attributed with the creation of Japan and of the gods in Japan (somewhat similar to Adam and Eve in the Bible).[20]

Izanagi and Izanami were actually siblings, and their son Hiruko was born with deformities of the arms and legs (malformation of bones) and could not stand up by himself even when he was three years old (two years old in Western age counting). He was thus abandoned and was cast into a boat at sea. That is all that is said of Hiruko in *Kojiki* and *Nihon-shoki*. Neither text records anything of the whereabouts of Hiruko afterwards.[21]

Then, during the fifteenth century, the story of Hiruko revived as a folklore legend, in which it was believed that he grew to be a strong man and drifted ashore in Nishinomiya in current Hyōgo prefecture. There he was worshiped at Nishinomiya Shrine as a god of the sea, called the Ebisu (*lit.*, "someone from outside" or "foreigner") God. Thus, the Hiruko God was resurrected as the Ebisu God. The worship of the Ebisu God derives from Japanese fishing communities regarding as sacred things that drifted ashore and also rocks found in the ocean. Fishermen occasionally found whales that drifted into their local bays or those that were stranded ashore. Because whales were the largest things that drifted ashore, fishermen worshiped whales as the Ebisu Grand God.[22]

Another reason for this derives from the fact that whales and large fish, such as yellowtails and bonitos, shared the feeding spots of smaller fish; thereby a pod of whales helped fishermen to locate a school of large fish. Thus the Ebisu God was revered as the God of Fisheries, who is usually depicted as a big chubby figure wearing a casual hunting kimono and holding a fishing pole in its right hand and a big yellowtail (or a sea bream) in its left. For the same reason, the Ainu people, the native ethnic minority group (like the Native American Indians in North America) now living in Japan's northernmost island, Hokkaidō, also regarded whales as one of the gods in their own mythology.[23]

Prior to the age of modern science, the Japanese were highly superstitious in general, but fishermen were more so because their lives were at the mercy of the vagaries of the weather. Fishermen and whalers worshiped the Ebisu God by regularly making offerings to their local Ebisu Shrines or Hiruko Shrines through donations of stone *torii* (sacred entrance gates) or stone lanterns. They also donated giant *ema* (*lit.*, "picture horse"; originally a wooden plaque with a colorful picture of a horse painted on it; it comes from the tradition where parishioners used to donate horses to shrines) with a picture of a whale painted on. By worshiping the Ebisu God, fishermen-whalers prayed for safe sailing and a good catch in the next season.[24]

In turn, because of their belief in whales as the Ebisu God, fishermen in some communities feared that the gods of the sea would punish them if they killed whales, and therefore they did not hunt them.[25]

Then, in the seventeenth century, with the rise of the power of merchants, the Ebisu God emerged as one of the household gods called the Seven Gods of Fortune. Thus, while fishermen worshiped the Ebisu God as a guardian for their safety at sea and for a good catch, merchants worshiped it for their business prosperity and for good fortune. (For instance, a brand of Japanese beer, called Yebisu Beer, is named after the Ebisu God and uses its face as the beer logo.) The worship of the Ebisu God became widespread among Japanese so that people refer to it as "Ebisu-sama" with the honorific suffix "sama." Consequently, there are about thirty-five hundred Ebisu Shrines in Japan that worship the Ebisu God, as their principal god. Among them, Nishinomiya Shrine in Hyōgo prefecture is designated as the supreme head shrine.[26]

For instance, surprisingly, there are at least as many as eighty-nine Ebisu Shrines in Ōita prefecture, Kyūshū, alone, out of which fifty-five have the name "Ebisu Shrine." The Ebisu God also became paired with another Seven Gods of Fortune, Daikoku-ten (God of Great Darkness or Earth and God of Wealth; a fusion of the Hindu god Shiva and the Shinto god Ōkuninushi). Representing the blessings of the sea (fisheries) and those of the earth (agriculture) respectively, statues of Ebisu-sama and Daikoku-ten are often housed side by side in shrines.[27]

Benevolence for Life and Death

To these beliefs in Japanese mythology and Shintoism, the Buddhist notion of benevolence for life and death was added. Buddhism was "imported" from China via Korea during the sixth century AD. Many renowned Japanese Buddhist monks in ancient times, including Saichō (September 767–June 822) and Kūkai (774–April 835), sailed to China to study Buddhism and founded their own schools of Buddhism, the Tendai School and the Shingon School, respectively, upon returning home. As a corollary, the Japanese infused a sense of benevolence for life and death into the sense of reverence for life, cherished all forms of life, and mourned their deaths. The Japanese extended benevolence to nonliving things and conducted funeral services even for tools when they became unrepairable.[28]

For instance, there is a memorial ritual for broken or rusted needles called *hari-kuyō*. Instead of disposing of them, Japanese bring "dead" needles to a nearby shrine or a temple, where they stick needles into tofu or *kon'nyaku* (a somewhat jelly-like yam cake), so that needles would be laid to rest in a soft "coffin" (because needles have been thrust against hard clothes while they were "alive"). This memorial rite is the Japanese way to express their appreciation to the needles that enabled them to sew kimonos and futons. This practice has been continued to this day.[29]

This notion of benevolence for life and death, as well as reverence for life, is the foundation of the Japanese culture of mourning the deaths of whales. It should be noted here that this belief system was not held exclusively by Asians, but was embraced by Westerners. As a prime example, through helping sick people and observing nature in Gabon, then a part of French Equatorial Africa, Albert Schweitzer (January 1875–September 1965) came to believe that the lives of all living things, from blades of grass to elephants, deserved to be equally respected and that reverence for life must be the guiding moral principle for human beings. He put this notion into practice in his own life and did not kill mosquitoes, showing benevolence toward the tiny insects.[30]

Syncretism of Shintoism and Buddhism

With the influence of both Shintoism and Buddhism, the Japanese at large—with the exception of genuine practitioners of Shintoism, who are in the minority—have worshiped both indigenous Shinto gods and deities in Buddhism such as Amida-Buddha and *Kan'non-bosatsu* (the Goddess of Mercy, one of the guardian deities of Amida-Buddha). The two beliefs have coexisted in the hearts of the Japanese and there were no contradictions between them. Most of the Japanese were secular practitioners of religions, and lay believers did not concern themselves with doctrinal precepts, but simply worshipped various deities—the more, the better.[31]

Consequently, a Shinto shrine strangely often exists right next to a Buddhist temple, and the Japanese visit both places on religious days or festivals. Also, a typical Japanese household has two altars in the best room of the house—one for Shintoism (*kamidana*) and another for Buddhism (*butsudan*)—and offers prayers at both every day. In addition, in the household worship of various deities over the course of time, gods in Shintoism and Buddhism (and Hinduism) were mixed together in Japan, and as a result, a curious syncretism of Shintoism and Buddhism occurred.[32]

A prime example is one of the Japanese goddesses in Buddhism called Benzaiten (or "Benten" in short), which is also one of the Seven Gods of Fortune, along with the Ebisu God. Benzaiten originated from the Hindu goddess Saraswati (a river deity, and a goddess of knowledge, music and arts) and was adopted by Buddhism. However, Benzaiten has also been worshiped at Shinto shrines, and Benten shrines became ubiquitous features of towns and villages throughout Japan. Benten Shrines were usually built on islands because Benten was originally a river goddess. A notable example is Itsukushima Shrine on Miyajima in Hiroshima prefecture. It is designated as a National Treasure of Japan and also is registered as a World Heritage Site by the United Nations Educational, Scientific and Cultural Organization (UNESCO).[33]

The worship of Benten had deeply penetrated into Japanese daily life so that it is referred to as "Benten-sama," with the honorific suffix. There is even a folklore legend about whales making a pilgrimage to a local Benten Island (see Chap. 9). Also, the fact that the Ebisu God is the only indigenous Japanese god of the Seven Gods of Fortune, whereas the other six gods were originally deities in Hinduism, Daoism, or Buddhism, indicates the eclectic adaptability of the Japanese, as well as the unusual fusion of Shintoism and Buddhism in Japan.[34]

Suppression of Buddhism

However, an immediate caveat should be given about this syncretism. The Shinto-centric school of thought emerged during the Edo period (1603–1868) among the scholars who advocated the study of things purely Japanese in the face of the teachings of Buddhism, which was a foreign religion imported from China. As a result, many Buddhist temples were abolished by the lords of provinces in certain regions. This movement was intensified and spread nationwide by the Meiji government, which was established in 1868 by restoring the power of the emperor, who was considered the descendant of the gods of Japanese ancient mythology and the head priest of Shinto. The government enacted policy abolishing the syncretism of Shintoism and Buddhism, and separating the two religions. This escalated into the so-called *haibutsu kishaku* (*lit.*, "abolish Buddhism, destroy Buddha"), which destroyed countless Buddhist temples, along with their sacred statues of Buddha and other cultural properties, and burned their sutras.[35]

Consequently, many Buddhist temples were either demolished or turned into Shinto shrines, and Buddhist monks were forced to convert to Shinto priests or suffered worse fates. Reflecting this complex history of "mergers and acquisitions" of Buddhist temples by Shinto shrines throughout Japan, many Shinto shrines absorbed some elements of Buddhism, complicating the syncretism of the two religions.[36]

Division of Labor Between Shintoism and Buddhism

Meanwhile, a general "division of labor" between Shintoism and Buddhism has developed in conducting secular ceremonies in the daily lives of the Japanese. Shinto is used to celebrate the birth and growth of children whereas Buddhism is used to mourn the death of people. For instance, a family takes a newborn baby to a Shinto shrine, like a christening ceremony at a Christian church. The family also takes their children to a Shinto shrine for the rites of passage at three, five, and seven years of age in the old age counting system called the *Shichi-go-san* (*lit.*, "seven-five-three").[37]

In addition, most of the Japanese have wedding ceremonies according to Shinto rituals. They used to have such ceremonies at home in the Shinto style, but the custom of having them at Shinto shrines began in 1900 when the Tokyo Grand Shrine conducted the Crown Prince's wedding ceremony. In turn, Buddhism serves to mourn the deaths of people and to conduct funeral services and subsequent periodic and seasonal memorial services. (As an exception, genuine Shintoists have their funeral services in the Shinto style.)[38]

A similar division of labor is observed in the way the Japanese dealt with whales. Generally speaking, Shinto shrines were the place to worship the whale god (the Ebisu God) and to offer prayers for safe sailing and a good catch, whereas Buddhist temples were the place to mourn the deaths of whales and to pray for the souls of whales so that they would attain the enlightenment of Buddha. However, due to the syncretism of Shintoism and Buddhism, and also due to suppression of Buddhism in favor of Shintoism, Japanese fishermen and whalers in some communities created graves for whales and erected monuments for whales at local shrines.[39]

In contrast, Buddhist temples were never used as the place for worshiping the Ebisu God, which was under the exclusive jurisdiction of Shinto shrines. Instead, Buddhist temples dealt with issues concerning how to take care of the souls of the dead whales and their "afterlives." Judging from the available records, it is safe to state that Shinto shrines assumed the primary role of the place to worship the whale god and offer prayers for safety of crewmen and a good catch, but also assumed the secondary role of memorials for whales in dedication to the Ebisu God. In contrast, Buddhist temples assumed overall responsibilities for atoning for the sin of killing whales and for offering prayers for the souls of whales so that they would rest in peace in heaven.[40]

Japanese Culture of Mourning the Death of Fish

Another interesting aspect of the Japanese memorial rites for whales is that they are actually a part of the Japanese culture of mourning the death of fish. There was no concept of marine mammals in traditional Japanese society, and whales were considered the largest fish. In recent years, foreign observers have noted the uniqueness of the Japanese culture of mourning the death of fish caught for human consumption. This tradition is not shared by the Chinese despite the fact that the Chinese were mentors for the Japanese. China was "Greece and Rome" to the Japanese, and the

Japanese learned practically everything from China, from the written language to ways of thinking and behaving—Buddhism, Confucianism, and Daoism.[41]

Nevertheless, the Chinese anthropologist Wang Min was surprised to find various forms of monuments dedicated to fish in Japan, as well as to animals and items such as chopsticks and combs. There are also graves for crabs and shrimps, as well as for shellfish (including Japanese pearl oysters that were cultivated for pearls) and kelp and other seaweed. Wang was impressed with the empathy and sympathy the Japanese held for fish and other marine life. Wang, who grew up in China and has lived in Japan for years, finds no equivalent tradition in China. Wang instead contrasts these Japanese traditions to the epicurean culinary culture of China: "The Chinese eat everything that flies, except for airplanes. They eat everything that has legs, except for chairs." (The Chinese did not seriously engage in whaling and do not have a tradition of eating whale meat.)[42]

Wang's first encounter with a fish grave was a 10-foot-long gravestone for eels erected at Zuigan Temple in Matsushima, Miyagi prefecture. Her Japanese friend told her that there were many eel graves all over Japan and that eel dealers held funeral services for eels. It was only many years later that she realized that Japanese benevolence for the life and death for all living things led them to mourn the deaths of eels killed for human consumption.[43]

The notion of mourning the death of fish is also foreign to Russians. During a study tour of the Japanese fisheries industry, Soviet officials found a tall monument with the inscription "Memorial Tower for the Souls of Salmon" erected at a salmon farm in Miyako, Iwate prefecture. They wondered why the Japanese erected such a respectable monument for mere fish raised to be eaten. (Similarly, the American Kentucky Fried Chicken headquarters do not understand why their Japanese subsidiary performs memorial rites for chickens.) This was the Japanese way of expressing their benevolence for marine life sacrificed for human consumption and of trying to atone for their deeds. Therefore, the fishermen created graves and erected memorial stones for fish and other marine life, and held funeral and memorial services services for them and prayed for their souls.[44]

Amazingly, throughout Japan at least 630 graves and monuments for fish and other marine life—excluding mammals such as dolphins and whales—have been identified. In earlier times, the actual total number would have been substantially higher because many of them have decayed and disappeared.[45]

Even today, fishermen and traders at the Tsukiji Fish Market in Tokyo (the largest fish market in the world) offer prayers at Namiyoke Inari Shrine, located near the market, where many monuments, including those for monkfish and shrimp, and even for eggs that were used in making sushi, were erected at the shrine, and the shrine conducts seasonal memorial rites and festivals for them. Fish traders and restaurateurs also offer prayers at Benten-dō (Prayer House for Benten) on Benten Island in Shinobazu Pond in Ueno, where numerous monuments, including those for blowfish and softshell turtles, as well as for cooking knives, were erected.[46]

Tsukiji Fish Market was scheduled to be relocated to a new site in Toyosu (former site of Tokyo Gas Company) in November 2016; however, this plan stumbled due to the abrupt resignation of Tokyo Governor Masuzoe Yōichi in June 2016 for his embezzlement of the government budget. The new governor Koike Yuriko conducted a new series of soil tests for the new Toyosu site. The tests discovered high levels of toxic substances—the "leftover" (such as industrial waste) from the days of Tokyo Gas Company—that had not been detected in previous tests, suggesting manipulations of the previous tests in favor of the relocation. Thus, the relocation of Tsukiji Fish Market has been suspended as of May 2017.[47]

Whale as the Largest Fish

Before the age of modern science, the Japanese considered whales to be the largest fish—because they lived in the ocean and they did not have legs. In fact, the Chinese character (which the Japanese language also uses) for the word whale means "huge fish." The character is made of two parts: the *hen* part (the element on the left side, functioning as a semantic classifier for the word, or a radical) denotes "fish," while the *tsukuri* part (the element on the right side, functioning as a specifier of the word) denotes "very large." All the names for fish in Chinese characters have the same *hen* part, whereas words that denote four-legged animals have a different *hen*.[48]

Meanwhile, the traditional mores of Japanese society were regulated by Buddhism, which proscribed killing living things in general and four-legged animals in particular. Thus, the Japanese did not eat four-legged animals, but eating fish was condoned, and fish constituted the main source of animal protein for the Japanese. Therefore, catching and consuming whales did not violate the code of Buddhism. It was not until the Meiji period (1868–1912), in which the new government promoted Westernization, that the Japanese began to eat beef and pork. In addition, while the

Japanese ate fish, they still refrained from eating fish on memorial days and religious days. This tradition is mostly kept in practice today.[49]

Japanese fishermen and whalers also had a deep sense of reverence for whales because they were mighty and magnificent. As fishermen-whalers hunted for whales, risking their own lives, they were in awe of the power and intelligence of whales. They also held a deep sense of appreciation for whales that had sustained their living, and they utilized the entire whale, not wasting anything. At the same time, they feared punishment from the gods of the sea for killing whales. Thus, fishermen-whalers feared and revered whales as the god of the sea simultaneously. Because seasonal whaling was a way of life for many fishing villages in Japan, the fishermen-whalers were confronted with the dilemma between their need to sustain their own livelihoods and their respect and sympathy toward the largest fish they had ever known.[50]

Japanese fishermen solemnly accepted their fate as whalers and sought to ease the wrath of the spirits of whales by praying for the souls of whales. They sought for atonement through religious rituals. They commissioned local temple head priests to perform various forms of memorial rites, normally given only to human beings, in which all the villagers in the community took part. In so doing, each community conducted memorial rites for whales according to the mores of that locale and the precepts of the specific school of Buddhism of the locale, but the underlying concept and sentiments were the same. Fishermen and whalers did not concern themselves with the doctrinal details of different sects of Buddhism, but only with how they could atone for their killing whales and how the souls of whales could rest in peace in heaven. This resulted in the Japanese culture of mourning the deaths of whales, which is one of the most unique features of Japanese whaling culture.[51]

* * *

Arrival of American Modern Whalers

The 1820s saw American and European whalers begin operations in the whaling grounds surrounding Japan, which drastically changed the lives of Japanese fishermen engaged in coastal whaling. Specifically, Americans sailing on whaleships manned by African-American laborers actively launched their operations in the Pacific Ocean, near the Izu and Ogasawara Islands (Japan

Ground) and off the Kamchatka peninsula (Kamchatka Ground). The American invention of gun harpoons and bomb lances in 1846 only accelerated their operations. Most of the American whaling bases were located in New England, and one of them, Nantucket, Massachusetts, flourished to the extent that Herman Melville wrote in *Moby Dick*, "two thirds of this terraqueous globe are the Nantucketer's. For the sea is his; he owns it, as Emperors own empires[.]"[52]

Principal characters in *Moby Dick*, such as Ahab and Starbuck, were from Nantucket. Thus, Americans developed their own whaling culture, as exemplified by scrimshaw—intricate carvings and engravings on the bone and baleen of whales, created as a traditional shipboard pastime by whalers. Americans hunted whales primarily for whale oil and therefore took the blubber of whales, but did not take whale meat for trade. They also took baleen to make such goods as buggy whips, corset stays, and umbrella ribs. Being used as fuel for lamps and lubricant for machines, whale oil was an important commodity for Americans. Whale oil even became a target of a New York City court case in 1818 concerning whether it was fish oil and therefore subject to state inspection and taxation, as chronicled in *Trying Leviathan: The Nineteenth-Century New York Court Case That Put the Whale on Trial and Challenged the Order of Nature* (2007) by D. Graham Burnett. By the mid-nineteenth century, the prosperity of Nantucket was replaced by that of New Bedford, Massachusetts, which is still referred to as "The Whaling City" because of this history.[53]

American Whaleship, John Howland

It is also interesting to observe that the volume of American whaleships was such that they often encountered Japanese shippers and that they even rescued shipwrecked Japanese fishermen in the Pacific Ocean. In one well-known case, the American whaleship John Howland rescued five Japanese fishermen who had drifted onto Tori Island in the waters off the Izu Islands in the Pacific Ocean in 1841. Nevertheless, the *sakoku* (seclusion) decree of the Tokugawa shogunate government (1603–1868) prohibited foreign ships from entering Japanese bays, so Japanese castaways were unable to be repatriated by foreign ships. In turn, for the Japanese, leaving Japan was an offense punishable by death. Captain William Whitfield thus took them on the voyage and left four adult fishermen in Honolulu, because they did not wish to go further away from home.[54]

By contrast, a fourteen-year-old boy stayed with the ship and landed in Fairhaven, Massachusetts. This illiterate young Japanese studied English and navigation there and eventually returned home in 1851. This is the famous story of Nakahama Manjirō (January 1827–November 1898) or "John" Manjirō ("John" Mung in English, after the whaleship that rescued him). Nakahama made a significant contribution to the conclusion of the US- Japan Treaty of Peace and Amity of 1854 and the US- Japan Treaty of Amity and Commerce of 1858, as an interpreter.[55]

American Whaleship, Manhattan

Subsequently, in March 1845, another American whaleship, Manhattan, rescued twenty-two shipwrecked Japanese, who had been adrift near Tori Island. Because foreign ships could not enter Japanese bays, Captain Mercator Cooper first sent four of the rescued Japanese to Tokyo Bay and had them deliver a message to the Tokugawa government, seeking permission to return the remaining Japanese. The permission was granted in April 1845, and the Manhattan became the first American ship that made an authorized visit to Tokyo Bay. While the ship was met with curiosity and hostility by Edoites at large, Tokugawa government officials were courteous. The government gave Captain Cooper generous amount of firewood, food, and water, but refused payment for the provisions.[56]

Meanwhile, in the wrecked Japanese ship, Captain Cooper found a map that charted the islands of Japan and handed it over to the US government upon returning. More than a century later, in July 1972, three Rotary Clubs in Tokyo erected a monument for the Manhattan in commemoration of this humanitarian act.[57]

Arrival of American Commodore Matthew C. Perry

Eight years after Captain Cooper's visit, American commodore Matthew C. Perry (April 1794–March 1858, the "uncle" five generations removed from former defense secretary William J. Perry in the Bill Clinton administration), visited Uraga (in current Kanagawa prefecture) in July 1853, with four warships, threatening the Tokugawa shogunate government into accepting an open-door policy ("Gunboat Diplomacy"). It is said that Perry used the map Captain Cooper took from the wrecked Japanese ship in 1845. Perry returned to Japan the following year, anchored off Yokohama, and succeeded in opening Japan by concluding the US- Japan Treaty of Peace and Amity of March 1854. It is important to note that one of the major reasons for Perry's visits to Japan in 1853 and 1854 was to arrange for firewood and charcoal for American whaleships and food and water for crewmen, as well as for means to repair the whaleships and rescue the injured crewmen.[58]

Russian and Norwegian Whaling Operations

Meanwhile, the Russian Pacific Whaling Company engaged in whaling grounds in the Sea of Japan. These operations by American and Russian whalers depleted whaling grounds in the Pacific Ocean and the Sea of Japan. Consequently, whales stopped coming to the shores of Japan by the 1860s. Norwegian whalers further revolutionized whaling methods by inventing cannon-fired harpoons. Employing steam-powered catcher boats equipped with bow-chaser deck cannons and heavy-caliber harpoons that exploded on impact, Norwegians actively engaged in whaling in the Atlantic Ocean and elsewhere.[59]

Curiously, in 1912, a young official at the Japanese Ministry of Agriculture and Commerce's Fisheries Bureau, Takasaki Tatsunosuke (February 1885–February 1964), who was working on loan to a Mexican fisheries company in Baja California, had a chance meeting with a Norwegian whaling ship. He observed in the lagoons of Magdalena Bay that Norwegian whalers took only the blubber of whales (for whale oil) and sailed off, leaving scores of whale carcasses on the shore. He described the scene as horrific, as well as sorrowful and wasteful. The scene Takasaki saw resonates with the scene of countless bodies of American buffalos left on the plains after hunters had skinned them for hides, as depicted in the Hollywood movie *Dancing with Wolves*, released in 1990.[60]

Meanwhile, Japanese whalers erroneously believed that the disappearance of whales was punishment from the gods of the sea (they thought that they had overhunted their prey) and made extra donations to shrines and offered extra prayers for the return of whales. Their prayers were not answered and Japanese classical whaling completely ended by the early twentieth century. Today, the old equipment for classical whaling is designated as tangible cultural properties in Japan, while memorial rites for whales and other forms of whaling culture are designated as intangible cultural properties.[61]

Kaneko Misuzu, a native of Senzaki, Yamaguchi prefecture, described the "glory" and sorrow of a former whaling bay village in this poem:

"Kujira-tori" (Whaling)
At night when the sea roared, in the winter night,
while roasting the chestnuts,
I listened to the story.
The story of whaling long, long ago,
here in this very bay, Shizuga-ura.
The sea was raging, the season was winter,
snowflakes blowing in the wind,
harpoon ropes flying in the snow.
The purple rocks and pebbles,
and even the water that was usually purple,
but even the shore turned red.
Wearing a double layer of padded kimono coats,
standing at the bow of the boat and watching,
as soon as the whale became weak,
they took off their clothes at once and dived naked into the raging waves.
They were the fishermen long, long ago—
listening to the story, my heart was also excited.
Whales stopped coming to the bay,
and the bay was impoverished.
The sea roars, in the winter night,
when the story ended and I noticed it—.[62]

Notes

1. Boreal Institute for Northern Studies, ed., *Small-Type Coastal Whaling in Japan: Report of an International Workshop*, Edmonton, Alberta: Boreal Institute for Northern Studies, the University of Alberta, 1988, 10–11.
2. Ibid.
3. Ibid., 11–12; Yoshihara Tomokichi, "Kujira no haka" (Graves of Whales), in Tanigawa Ken'ichi, ed., *Nihon minzoku bunka shiryō shūsei* (Collection of Documents on Japanese Folk Culture), Vol. 18, Tokyo: San'ichi-shobō, 1997, 433, 469.
4. Ibid. (both).
5. Kumano Taiji-ura hogei-shi hensan-iinkai, ed., *Kumano no Taiji: Kujira ni idomu machi* (Taiji, Kumano: Town That Challenged Whales), Tokyo: Heibonsha, 1965, 5–15.

6. Ibid., 7–15; Boreal Institute for Northern Studies, 10–11.
7. Ibid. (both).
8. Kumano Taiji-ura hogei-shi hensan-iinkai, 24–25, 88–90.
9. Ibid.
10. Boreal Institute for Northern Studies, 79–84; Nobuhiro Kishigami, "Hogei ni kansuru bunka-jinruigaku teki kenkyū ni okeru saikin no dōkō ni tsuite" (Recent Trends in Cultural Anthropological Studies on Whaling), Kokuritsu minzoku hakubutsukan kenkyū-hōkoku, 2001, Vol. 35 No. 3, 399–470, http://ir.minpaku.ac.jp/dspace/bitstream/10502/4475/1/KH_035_3_001.pdf, downloaded February 28, 2016.
11. Ibid. (both).
12. Ibid. (both); Kumano Taiji-ura hogei-shi hensan-iinkai, 91–96.
13. Boreal Institute for Northern Studies, 53–59; Yoshihara, 411–421, 436–440, 449–450.
14. Ibid. (both).
15. Ibid. (both).
16. Ibid. (both).
17. Ibid. (both).
18. Wang Min, *Nitchū 2000-nen no furikai: Kotonaru bunka "kisō" o saguru* (Two Thousands Years of Misunderstanding Between China and Japan: To Grope for "Fundamental Stratum" of Cultural Difference), Tokyo: Asahi Shinsho, 2006, 82–84, 179–184. For detail, see Nakamura Ikuo, *Saishi to kugi: Nihonjin no shizenkan · dōbutsu-kan* (Religious Dedications and Sacrifices: Japanese Views of Nature and Animals), Kyoto: Hōzōkan, 2001 and *Nihonjin no shūkyō to dōbutsu-kan: Sessho to nikujiki* (Religion and Views of Animals of the Japanese: Killing and Meat-eating), Tokyo: Yoshikawa-Kōbunkan, 2010; Pamela J. Asquith and Arne Kalland, eds., *Japanese Images of Nature: Cultural Perspectives*, Richmond, Surrey: Curzon Press, 1997.
19. Ibid.
20. Harada Shin'ichi, "Hirugo [*sic*] shinwa ronkō" (Analysis of Hirugo Mythology), March 1994, http://repo.komazawa-u.ac.jp/opac/repository/all/16167/KJ00005086929.pdf, 147–157 and 179–180.
21. Ibid.
22. Ibid., 149; Kumano Taiji-ura hogei-shi hensan-iinkai, 17–18.
23. Ibid. (both).
24. Harada, 149; Boreal Institute for Northern Studies, 59–64; Yoshihara, 411–420.
25. Boreal Institute for Northern Studies, 59–64; Yoshihara, 411–420.
26. Harada, 149–151; "Zenkoku Ebisu-ōkami hōsai-sha ichiran" (List of Shrines in Japan that Worship Ebisu Grand God), http://nishinomiyaebisu.web.fc2.com/todouhuken/menu/table/43table.html, accessed January 18, 2016.

27. Ibid. (both).
28. Boreal Institute for Northern Studies, 52–53; Wang, 59–65. For detail, see Nakamura Hajime, *Daijō Bukkyō no shisō* (Philosophy of Mahayana Buddhism), Tokyo: Shunjūsha, 1995 and Nakamura Hajime, *Jihi* (Benevolence), Tokyo: Kōdansha, 2010; Tan'no Akira, *Kurashi ni ikiru Nihon no shikitari* (Customs in Japanese Daily Life), Tokyo: Kōdanshia, 2000.
29. Angelika Kretschmer, "Mortuary Rites for Inanimate Objects: The Case for Hari Kuyō," *Japanese Journal of Religious Studies*, Vol. 27, No. 3/4, Fall 2000, 379–404.
30. Boreal Institute for Northern Studies, 52–53. For detail, see A. G. Rud, *Albert Schweitzer's Legacy for Education: Reverence for Life*, New York: Palgrave Macmillan, 2011.
31. Ibid. For detail, see Nakamura Hajime, *Ways of Thinking of Eastern Peoples: India, China, Tibet, Japan*, trans., Philip P. Wiener, Honolulu: East-West Center Press, 1964; Nakamura Ikuo, *Kami to hito no seishin-shi: Nihon Bukkyō no shinsō-kōzō* (Spiritual History of Gods and People: Deep Core Structure of Japanese Buddhism), Kyoto: Jinbun-shoin, 1988.
32. Ibid.; Harada, 149–151.
33. Barbara R. Ambros, *Bones of Contention: Animals and Religion in Contemporary Japan*, Honolulu: University of Hawai'i Press, 2012, 76–80; "Itsukushima Shinto Shrine," http://whc.unesco.org/en/list/776, accessed March 6, 2016. For background, see Mark Teeuwen and Fabio Rambelli, eds., *Buddhas and Kami in Japan: Honji Suijaku as Combinatory Paradigm*, New York: RoutledgeCurzon, 2003.
34. Harada, 149–151; Yoshihara, 473.
35. Boreal Institute for Northern Studies, 59–61. For background, see Teeuwen and Rambelli.
36. Ibid.
37. Ibid., 52–53. For background, see John Breen and Mark Teeuwen, eds., *Shinto in History: Ways of the Kami*, Honolulu: University of Hawai'i Press, 2000 and *A New History of Shinto*, Malden, MA: Wiley-Blackwell, 2010.
38. Ibid.
39. Ibid., 53–64; Yoshihara, 410–421. Due to the sheer number of Ebisu Shrines in Japan—thirty-five hundred in total—this book examines only those shrines that house whale graves and/or monuments.
40. Ibid. (both).
41. Yoshihara, 410–411; Wang, 63–66, 82–84.
42. Ibid. (both).
43. Wang, 64–66; Kumano Taiji ura hogei-shi hensan-iinkai, 83–84.
44. Yoshihara, 410–411.
45. Ibid.

46. Ambros, 76–79; "Tsukiji Namiyoke Jinja," http://www.namiyoke.or.jp, accessed March 12, 2016.
47. "Toyosu no giron 'Hayaku saikai o'" ("Resume Quickly" the Debate on Toyosu), May 19, 2017.
48. Boreal Institute for Northern Studies, 66–70. For detail, see Nakamura Ikuo (2010).
49. Ibid.
50. Ibid., 52–62; Hosokawa Takao, "Kujira ni 'kansha to ikei' no nen" (Feelings of "Appreciation and Respect" for Whales), *Mainichi Shimbun*, October 26, 2010.
51. Boreal Institute for Northern Studies, 52–65.
52. "American Whaling," http://www.whalingmuseum.org/learn/research-topics/overview-of-north-american-whaling/american-whaling, accessed January 14, 2016; Herman Melville, *Moby Dick* or, THE WHALE, New York: Harper & Row, 1966 (originally published in 1851), 56.
53. "American Whaling"; D. Graham Burnett, *Trying Leviathan: The Nineteenth-Century New York Court Case That Put the Whale on Trial and Challenged the Order of Nature,* Princeton, N.J.: Princeton University Press, 2007.
54. Yoshihara, 425–426; Kumano Taiji-ura hogei-shi hensan-iinkai, 25. For detail, see Donald R. Bernard, *The Life and Times of John Manjiro*, New York: MacGraw-Hill, 1992.
55. Ibid. (both).
56. Yoshihara, 425–427.
57. Ibid.
58. Ibid., 426; Kumano Taiji-ura hogei-shi hensan-iinkai, 144–145; Komatsu Masayuki, *Yutakana Tokyo-wan: Yomigaere Edo-mae no umi to shoku-bunka* (Rich Tokyo Bay: Restore the Sea and Food Culture of Edo), Tokyo: Yūzankaku, 2007, 91–95.
59. Kumano Taiji-ura hogei-shi hensan-iinkai, 160–164.
60. Takasaki Tatsunosuke, "Watashi no rirekisho" (My Autobiography), in Nihon keizai shimbunsha, ed., *Watashi no rirekisho* (My Autobiography), Vol. 1., Tokyo: Nihon keizai shimbunsha, 1980, 436–437.
61. Kumano Taiji-ura hogei-shi hensan-iinkai, 145–146.
62. Kaneko Misuzu, *Kaneko Misuzu zenshū* (Complete Works of Kaneko Misuzu). Vol. 3. Tokyo: JULA shuppan-kyoku, 1984, 201–202.

CHAPTER 3

Whaling Culture and Whale Graves in the Hokkaidō and Tōhoku Regions

Classical Whaling in Hokkaidō

Hokkaidō, the northernmost island of Japan, was geographically and culturally separated from the Japanese main island, Honshū, and was referred to as *Ezo-chi* (Ezo land) until 1869. It had been sparsely populated, and mostly the Ainu people (an ethnic minority group in Japan, similar to indigenous peoples in North America) lived there. Unlike the Japanese (the Yamato people), who were primarily farmers and mainly lived by growing rice and vegetables, the Ainu people were primarily hunters and mainly lived by hunting animals (such as Ezo brown bears and Ezo deer) as well as collecting fruits and plants. There are few records of classical whaling in Hokkaidō, because the Ainu people did not have a written language. In 1945 the first book about Ainu whaling was written, *Funka-wan Ainu no hogei* (Ainu Whaling in Funka Bay), by Natori Takemitsu.[1]

Only in recent years has the oral history of Ainu whaling been compiled through interviews with Ainu elders, who remember what they heard from their ancestors. This was part of a project to preserve the Ainu spoken language called the Hokkaidō Ainu, which was designated as a "critically endangered language" by UNESCO in February 2009. The Ainu people referred to whales as "old men offshore" and revered them as the "gods that bring fish to the shore." This belief came from the fact that a pod of whales and a stock of Pacific herring or a run of Pacific salmon shared the feeding

M. Itoh, *The Japanese Culture of Mourning Whales*,
https://doi.org/10.1007/978-981-10-6671-9_3

spot of smaller fish and thereby enabled the Ainu fishermen to locate these fish.[2]

While revering whales as gods, the Ainu people caught whales that were chased by orcas (killer whales) and stranded ashore, and those that ran aground in the drift ice, and consumed them as blessings of the gods of the sea. The Ainu people also sailed out to hunt whales, applying the same hunting methods that they employed in the mountains: using arrows coated with a fatal toxin, aconitine, that was extracted from aconite (also known as devil's helmet or monkshood). Usually, five men sailed out on a small boat and shot the arrows into a whale. They sailed back immediately, before the whale became violent. Then they waited several days or more until the poisoned whale washed ashore. Each arrow had a family mark of an individual hunter so that whoever found the whale could tell who had shot it. If the person who found the whale was from a different community, the hunter's community gave part of the whale meat to the community of the finder. Then the hunter's community shared the rest of the whale meat with all its members.[3]

Some fishing communities in Funka Bay (also called Uchiura Bay) and Hakodate Bay employed a more active method of whaling by using aconitine-coated manual harpoons, because these enclosed bay areas made it easier to catch whales there. In this method, the heads of the poisoned harpoons were detachable so that they would stick in the body of a whale, and when the whale became weak, the hunters pulled on the harpoon ropes (they did not have to wait until the poisoned whale washed ashore). The aconitine method of Ainu whaling ended in 1871 when the Meiji government banned all hunting using poisoned spears throughout Japan.[4]

The oral history of the Ainu people does not record anything about whale graves or monuments. They apparently did not conduct memorial rites for whales. This makes sense because the Japanese culture of mourning the deaths of whales derived from Buddhism, and the Ainu people did not subscribe to Buddhism. Instead, the Ainu people offered prayers to the whale gods before and after hunting, as the Japanese (the Yamato people) did at Shinto shrines. The Ainu people also dedicated "whale dances" to the whale gods. Each community has its own versions, such as *Kiritappu-hama drum* in Hamanaka and the whale dance and whale festival in Shiraoi-chō. They are only part of the rich Ainu folklore culture. These dances have been preserved and are presented today at local "whale festivals."[5]

Age of Modern Whaling in Hokkaidō

Meanwhile, in March 1854, the Tokugawa shogunate government, which had enforced the *sakoku* (seclusion) policy for more than two centuries, opened the country to the outside world by concluding the US-Japan Treaty of Peace and Amity (or the Kanagawa Treaty, after the site of the convention). As a result, the Shimoda port on the southern tip of the Izu peninsula and the Hakodate port on the southern tip of Hokkaidō were opened to American vessels. Russia followed suit and concluded the Russo-Japan Treaty of Commerce and Navigation (or the Shimoda Treaty) in February 1855, opening the ports of Shimoda, Hakodate, and Nagasaki to Russian vessels. Accordingly, the Hakodate port was open to American and Russian vessels, both of which were engaged in modern whaling in the Pacific Ocean.[6]

There is a fine three-folder *Nishiki-e* (multi-colored woodblock print) of the Hakodate port around 1868 (artist unknown) depicting the buildings with the names of America, England, and Russia, as well as American whaleships in the bay. In this context, the Tokugawa shogunate government in 1857 sent Nakahama Manjirō (or John Mung, see Chap. 2) to Hakodate and had him teach modern American whaling methods to local fishermen, without much success.[7]

Then, things changed drastically when the Meiji government took power from the Tokugawa government in 1868. As soon as the Meiji government renamed *Ezo-chi* to Hokkaidō and incorporated it into its jurisdiction in 1869, it promoted development and settlement projects for the northern frontier on Hokkaidō, along with an enthusiastic drive for modernization and Westernization of Japan. Consequently, Hakodate became one of the modern whaling bases in Japan at the turn of the twentieth century. In addition, in 1915, the Tōyō-hogei Company opened a branch office in Abashiri (a phonetic rendering of the Ainu name of the locale into the Japanese alphabet) in the northeast of Hokkaidō facing the Sea of Okhotsk, and began modern whaling there.[8]

Roles of Religion in Whaling

Religion has also played an important role in modern whaling in Hokkaidō. For instance, a new shrine was founded in Abashiri in 1908 through the process of *bunrei-kanjō* (division and propagation of the "souls of the gods"; like propagation of plants by dividing the roots) from Itsukushima

Shrine in Hiroshima prefecture. Then, at the beginning of each whaling season, a head priest of Abashiri Shrine conducted a ceremony on the whaleboat in order to purify the boat and the crewmen so that they were pure enough to be accepted by the gods of the sea. Crewmen then made offerings of a sacred tree (*masakaki*), salt, rice, and water to the small altar created in the whaleboat, and prayed for safe sailing and a good catch. Their wives meanwhile prayed to the gods for the safety of the crewmen by visiting the shrine every day.[9]

This ritual is in line with the tradition of the Japanese commissioning a Shinto priest to perform a purification ceremony of the land before building a new house or a new building. Also, just as every Japanese household has an altar for the Shinto gods (*kami-dana*), whalers created small shrines at their processing factories and offered prayers to the gods of the sea every day. Other whaling communities in Japan have conducted similar purification rituals.[10]

Among the crewmen of whaleboats, harpooners assumed a primary role in these rituals because they were most directly involved in killing whales. Once they caught a whale, their whaleboats encircled it and harpooners led crewmen in saying prayers for the whale. This is where Buddhism comes in. After the whaling season was over, the harpooners gathered to set a date for a funeral service for whales and commissioned their local temple head priest to perform it. They then regularly visited the local temple to pray for the souls of whales in person and also attended seasonal memorial services for whales until the next whaling season began. Thus, the religious rites concerning whaling complete a full cycle.[11]

Cenotaph for Whales in Hakodate

As a token of the early years of modern whaling in Hokkaidō, there is a respectable cenotaph for whales in Yayoi-chō, Hakodate (see Map and Table A.1). Former Enyō-hogei Company president, Amano Daisuke, who used to be a harpooner, erected this memorial in September 1957, at age 83. The main structure of about 7.9 feet in height and 7.5 feet in width is made of blocks of stone. In the center of this structure, a separate stone made of granite is attached, which has the inscription:

"*Kujira-zoku kuyō-tō*" (Memorial Cenotaph for Genus of Whales)[12]

In addition, a rectangular cube of granite stands on top of the main structure, which has inscriptions on all four sides. The inscription on the front side reads:

> I have caught more than two thousand whales since 1907. Reaching old age, I was overwhelmed with the sins of my business ... I erected this cenotaph in order to atone for my sins and to pray for the souls of whales so that they would attain the enlightenment of Buddha.[13]

Inscriptions on the other sides include a description of North Pacific right whales and of the local history of whaling. Further, on top of the rectangular cube sits a large replica of a North Pacific right whale. This monument suggests that modern Japanese whalers, much like their classical counterparts, had remorse and sympathy for whales, and tried to atone for their sins.[14]

Whale Temples and Shrines in the Northeastern Region

Moving down to the northeastern region of Honshū on the Pacific Ocean side, there are only sporadic records of classical whaling in fishing villages on the Sanriku Coast, comprising parts of Aomori prefecture, Iwate prefecture, and Miyagi prefecture. Adorned with many islands on the scenic coastline like jewels on the sea, the region was designated as the Rikuchū-kaigan National Park of Japan. Along with Fukushima prefecture, located to the south of Miyagi prefecture, the area is well known for the East Japan Massive Earthquake and Tsunami of March 11, 2011, where scores of cities and towns were almost totally destroyed. Two years after the disaster, in May 2013, the Rikuchū-kaigan National Park was renamed the Sanriku Fukkō (Reconstruction) National Park, in order to promote the reconstruction of the region.[15]

In fact, this author's visits to this area in August 2012 were extremely difficult, because most of the railroads were still cut off and the only public transportation available was the ad hoc and infrequent bus service run by private companies as a contingency measure.

There were no major classical whaling bases in the northeastern region of Honshū; however, the local history recorded that several communities had engaged in "passive whaling" during the eighteenth and nineteenth centuries and caught stray whales that had drifted into local bays. For instance, a classical whaler from Hizen-Ōmura province (in current Nagasaki prefecture), Ikeno Heijūrō, moved to Hachinohe province (roughly corresponding Aomori prefecture) and requested from the province lord a permit to engage in coastal whaling in 1699. (Classical whaling operations required a permit from a local province during the Edo period.) The province lord issued a seven-year permit to Ikeno beginning in 1700, in exchange for a sizable annual permit fee and other payments to the province.[16]

In addition, Kondō Den'emon, who was the biological son of Taiji Kakuemon, the legendary pioneer and inventor of the net method of classical whaling (see Chap. 6), engaged in whaling operations in current Ena-hama in Fukushima prefecture in the mid-eighteenth century. Trade of whales between Taiji (in Wakayama prefecture) and Edo (Tokyo) had prospered during the 1720s, and Taiji Kakuemon sent his son, Den'emon, to Edo as his agent to manage his trading business there. However, Den'emon embezzled his father's business funds. Having been disgraced, Den'emon left Edo and moved to Ena village. There he was adopted by a local samurai, who was a descendant of Satō Tadanobu (1161–1186), one of the "four guardians" of Minamoto Yoshitsune.[17]

Then, Den'emon began his whaling operations by introducing his father's net method. A local history records that he made a trade of whaling nets with Ōbi Gihee in 1775. His business was successful, so the local lord granted him privileges to assume a surname (Kondō) and to wear Japanese swords (these privileges were given only to men of the rank of samurai during the Edo period). Elsewhere, in Ōsuhama in Oshika county (in current Miyagi prefecture), a local member of the gentry, Abe Genzaemon, organized the first whaling guild in Oshika, and obtained a permit as well as financing from the local province lord in 1839. Nevertheless, he was only able to catch six whales in the spring season and his business failed. With this history of classical whaling, there are many monuments for whales on the Sanriku Coast.[18]

Whale Graves and Monuments in Iwate Prefecture

Memorial Stone for Steller Sea Lions in Ryōishi, Kamaishi

No whale graves or monuments have been preserved in the northern part of Iwate prefecture today, but there is a rare monument for Steller sea lions in Ryōishi, near Kamaishi, a major city in the central part of Iwate prefecture (see Map and Table A.1). The area was severely hit by the East Japan

Massive Earthquake and Tsunami of March 2011. The memorial stone for Steller sea lions, erected in 1858, stands at the entrance to a small shrine that houses stone statues of *Jizō-bosatsu* (Guardian Deity of Motherhood and Unborn Children), which sit next to fallen trees. This monument is made of a curved stone and withstood the 2011 Earthquake/Tsunami. The inscription on the monument reads: "*Todo kuyō-hi*" (Memorial Stone for Steller Sea Lions).[19]

The signboard next to the monument has a description of its provenance written by the Kamaishi Education Board in September 2004: "During the Edo period, the local fishermen had been troubled by colonies of Steller sea lions that migrated from the northern sea to Ryōishi Bay every winter. As Steller sea lions damaged the fishing grounds by devouring the fish, the fishermen caught them and shipped the meat and hides to Morioka, the capital of Morioka province. At the same time, the fishermen felt sorry for the Steller sea lions and erected a monument for them in March (April in the Gregorian calendar) 1858."[20]

Next to this monument stands a newer and larger memorial stone. This was erected for salmon by a local fisheries cooperative union leader in June 1989. The inscription on this monument reads: "*Sake kuyō*" (Memorial for Salmon). A separate auxiliary monument has an inscription of the provenance of the main monument written by a fisheries cooperative union board member.[21]

Whale Grave in Rikuzen-Takata

Going down from Kamaishi to the southern edge of Iwate prefecture, one arrives at Rikuzen-Takata. The city became internationally known as the epitome of the destruction caused by the East Japan Massive Earthquake and Tsunami, and the *Kiseki no ippon-matsu* ("Miracle Pine Tree") in the city that survived the disaster became its symbol. The Hirota peninsula on the southeastern tip of Rikuzen-Takata was separated from the mainland and became an instant island in March 2011. As with most of the buildings on the peninsula, the Hirota Post Office was swept away by the tsunami. The miracle pine tree stood by itself in August 2012, but the tree was soon declared dead due to immersion in seawater. A controversial restoration project for the tree began in September 2012, and the tree was given an internal support and artificial pine needles, so that it would forever look alive, as a monument of the hope for reconstruction.[22]

There is a whale grave in the district of Ōyō, Hirota-chō on the Hirota peninsula for a whale that was caught in January 1905. On the morning of January 24, an 82-foot-long blue whale that had been attacked by orcas swam into Hirota Bay. Local fishermen of both Ōyō and neighboring Tomari found the whale and competed to catch it. In the end, the fishermen of Ōyō won. The local government took about half of the profit of the whale and returned an equal share of the remainder to Ōyō and Tomari. Then, the fishermen created a grave for the whale. The gravestone is made of natural stone with a size of about 2.7 feet in height and 1.1 feet in width. It has the inscription, "*Tomurau ō-kujira no rei*" ("To Mourn the Soul of the Big Whale").[23]

It currently stands by the roadside on the coastline in front of a small shrine, along with several other monuments. It is flanked by a larger monument dedicated to the Great Dragon God on the left and a piece of the weathered whale bone and a stone statue of the Ebisu God on the right (see Photograph 3.1).[24]

Photograph 3.1 Whale grave (second from left), Hirota, Rikuzen-Takata, Iwate prefecture, 1905

Whale Temple in Rikuzen-Takata

In addition, there is a significant whale temple called *Kodate kan'non-dō* (*lit.*, "little prayer house for the Goddess of Mercy") in the district of Tomari, Hirota-chō. It houses a statue of "*Gyoran kan'non* (the Goddess of Mercy for Fish) Dedicated to Whales" ("the Goddess of Mercy for Whales" hereafter). The local people refer to the temple as "*Kujira kan'non*" ("Whale Kan'non" and hereafter) because the statue was created from the profit of a whale that had been stranded ashore. At the back of this small temple stands a stone statue of the Ebisu God. This is No. 25 Temple (the twenty-fifth station) of the Pilgrimage Loop of the "Thirty-three Goddesses of Mercy Temples in Kesen." In turn, the Goddess of Mercy for Fish is one of the "Thirty-three Goddesses of Mercy" in Mahayana Buddhism, and its statue usually takes the form of a beautiful woman carrying a basket of fish.[25]

This goddess is based on the folklore legend in Tang-dynasty China: A beautiful woman selling fish made a condition for marriage that the groom-to-be must become a religious subscriber of Hokke School Buddhism. She died soon after the marriage, and her husband realized that she was an embodiment of the Goddess of Mercy for Fish. Following this ancient story, Japanese fishermen in many communities made statues of the Goddess of Mercy for Fish and prayed for the souls of fish.[26]

However, this statue of the Goddess of Mercy for Whales in Tomari is unique because it takes the many-armed figure of *senju* (*lit.*, "thousand-arms"), which symbolizes many hands to help ease the sufferings of people. While there are many statues of the Thousand-armed Goddess of Mercy in Japan, including the famous one in Rengeō-in Sanjūsangen-dō in Kyoto, no other statue of the Thousand-armed Goddess of Mercy for fish or for whales exists. This statue calls for investigation because, despite its significance, Yoshihara Tomokichi, who conducted a most comprehensive study of whale graves and monuments in Japan, does not mention it. The local history tells us that the Whale Temple was founded by the head priest of Shingon School Buddhism Kaiun-zan Chifuku-in sometime during the Hōei period (April 1704–June 1711). This wooden temple on a hill withstood the 2011 Earthquake/Tsunami, but the search for this whale temple was an almost "stranger-than-fiction" story.[27]

In recent years, the local family Kumagai, who is related to the head priest of Rinzai School Zen Buddhism Kezō Temple in Otomo-chō, north of Tomari, has been the formal caretaker of the Whale Kan'non. Nevertheless, the Kumagai family does not actually manage the whale temple. Instead, on Kumagai's behalf, another Rinzai School Zen Buddhism temple, Jion Temple, located near Whale Kan'non, has managed this little temple. Through the good offices of Suda Keiko, who grew up in this community, this author was finally able to contact Jion Temple head priest Furuyama Keikō and also obtained Kumagai's permission for a private viewing of the statue. However, it turned out that Kumagai had only one of the two entrance keys of Whale Kan'non. Usui Tsutomu, who is a local fisheries cooperative union executive and president of the Service Society for the Statue of the Goddess of Mercy for Whales, had another entrance key.[28]

Worse, Usui's house was destroyed and washed away by the 2011 Earthquake/ Tsunami, and Usui had been evacuated to a shelter. Worse still, Usui had a stroke and was hospitalized. It thus became difficult to communicate with him and to locate the entrance key. Eventually, the Jion Temple head priest found out from the head priest of Rinzai School Zen Buddhism Kezō Temple that Usui's entrance key to Whale Kan'non had been washed away by the tsunami, along with his house. Kumagai then obtained Usui's permission to make a new key before this author's visit. This is how the rare viewing of the statue of the Thousand-armed Goddess of Mercy for Whales was made possible, with the attendance of a local historian and *Tōkai Shimpō* managing executive and editorial bureau chief Sasaki Katsutaka, as well as of Suda Keiko and others.[29]

The statue is gold-painted all around, with a large boat-shaped halo behind its back covering the whole figure, and is about three feet in height including the pedestal. It was carefully kept in a matching gold-painted, black-lacquered case (like a refined altar for human beings) and was tightly wrapped with cotton. It retained its golden sheen because it was repainted in 1965 when Whale Kan'non was rebuilt (see Photograph 3.2). On the back of the statue, there is an inscription dated September 19, 1965, indicating that Usui Yoshio (who could be Usui Tsutomu's father or grandfather) made a financial contribution to rebuild the temple in order to preserve the memorial rites for whales. It also has the name of Jion Temple twentieth head priest Shōdō-Tōkō (his secular family name was not given), who could be Furuyama Keikō's father or grandfather. In addition, there is a much older wooden (probably the original) statue of the Thousand-armed Goddess of Mercy for Whales, as well as a wooden statue of Amida-Buddha on the altar.[30]

The gold-painted statue of the Thousand-armed Goddess of Mercy for Whales had been open to the public only at the annual memorial service on the Autumnal Equinox, which is one of the important seasonal memorial days of Buddhism. For this ceremony, local fishermen carry the statue from Whale Kan'non down the hill to the Tomari fishing port, where the service

Photograph 3.2 New gold-painted statue of the thousand-armed goddess of mercy for whales, Kan'non-dō, Hirota, Iwate prefecture, circa 1965

is held. The memorial service for 2011 was cancelled due to the destruction of the area by the earthquake and tsunami. *Tōkai Shimpō* reported on the 2010 service as follows: "This year's memorial service was held yesterday, hosted by Jion Temple (head priest Furuyama Keikō), the Service Society for the Statue of the Goddess of Mercy for Whales (president Usui Tsutomu), and the Hirota Bay Fisheries Cooperative Union Hirota Branch (director Murakami Ichizō), which about seventy people attended."[31]

In the opening address, President Usui stated that he was grateful that he was able to see the statue again and to conduct a memorial service this year again. Director Murakami stated that fish and marine plants had sustained the local economy (the community stopped whaling a long time ago) and that he wholeheartedly prayed for the souls of the marine life, as well as for the blessings of the sea and the safety of fisheries operations in the next season. Then, while Jion Temple head priest Furuyama Keikō chanted the sutra, the attendants offered their prayers and burned incense.[32]

Whale Ema in Tomari, Rikuzen-Takata

Further, there is also a giant *ema* ("picture horse") of a whale at a Shinto shrine, Kakura-Myōjin Shrine, which the local people call "Okakura-sama," in the neighborhood of the Whale Temple. *Ema* is originally a wooden plaque with a picture of a horse painted on it, which parishioners donated to their shrine as a seasonal offering to the gods. This *ema* in Tomari is special because it has a picture of a stranded whale in the center, with two small fishing boats and fishermen. Local fishermen's head Ōwada donated the *ema* to their tutelary shrine, Kakura-Myōjin, in January 1891, in dedication to the god of whales. (There are two dozen whale *emas* preserved at shrines in other regions.) The existence of the rare statue of the Thousand-armed Goddess of Mercy for Whales and of this giant whale *ema* in Tomari, as well as of the whale grave in Hirota, suggests the depth of local people's benevolence for whales in Hirota Bay.[33]

Whale Shrines and Whale Temples in Miyagi Prefecture

Whale Shrine in Karakuwa, Kesen'numa

Going south from Rikuzen-Takata, and thereby entering Miyagi prefecture, there is a whale shrine called Osaki Shrine at the tip of the Karakuwa peninsula. It is located at Sakihama, Karakuwa-chō, which was incorporated into Kesen'numa city in March 2006. Local history records that in the early fourteenth century a member of the Ōji family, who was a descendant of the Itō clan—the local lord of Obi-Hyūga province (current Miyazaki prefecture) in Kyūshū—left his hometown, which had been in the middle of a civil war, in order to save the local shrine, Osaki Shrine, by transferring it to a safe, remote region. As Ōji and some of the Itō clan members sailed the ocean, their ships were guided by a white whale and arrived at the shore of Sakihama, Karakuwa village at the tip of the Karakuwa peninsula. They settled down there and built a new Osaki Shrine in 1308/1309.[34]

As examined in Chap. 2, whales have been revered as the Ebisu God (God of Fisheries) by the Japanese. Originally known as the Hiruko God in ancient Japanese mythology, the Ebisu God has become one of the household gods called the Seven Gods of Fortune and has been widely worshiped throughout Japan, along with Dakoku-ten (God of Great Darkness [Earth] or God of Wealth). Following this tradition, Osaki Shrine has a smaller shrine (separate from the main shrine) for the Ebisu God, which houses stone monuments of the Ebisu God and Dakoku-ten. In front of each monument sit colorfully painted new statues of the Ebisu God and Dakoku-ten.[35]

Whale Graves at Osaki Shrine in Karakuwa

Curiously, a group of three whale graves, made of stone, stand in the compounds of Osaki Shrine in Karakuwa-chō. The gravestone on the left side was made for a stray whale that drifted into the local bay and was caught in 1810. The inscription reads:

> "*Kujira-zuka*" (Whale Grave) with the name of the contributors "Fifty fishermen of Tadakoshi Bay, represented by Neguchi Yozaemon" and the date "April 3, Bunka 7". (May 5, 1810 in the Gregorian calendar)[36]

The local fishermen of Tadakoshi Bay, which is only five miles away from Osaki Shrine, felt sorry for the whale and made a financial contribution to the shrine in order to make this whale grave. The smaller grave on the right side was made for another stray whale that stranded in the Tadakoshi Bay in 1835. The inscription for this grave reads:

> "*Kujira-zuka*" (Whale Grave), "Tenpō 6, March 21". (April 18, 1835 in the Gregorian calendar)[37]

In turn, the largest gravestone in the middle is enigmatic. The upper part of the gravestone has a large inscription of three Chinese characters written in an archaic style. Some think that the three characters denote "whale," but this is not established. Rather, the three characters actually look like "Ba-shō-ō" (Venerable Bashō), referring to the famous haiku poet, Matsuo Bashō (1644–November 1694). Bashō did travel in northeastern Japan and wrote many haikus, some of which were compiled in his masterpiece journal, *Oku no hoso-michi* (*lit.*, "narrow streets of deep Japan"). This hypothesis is supported by the fact that the lower part of the gravestone has an inscription in three vertical lines written in cursive Japanese. This is consistent with the style of haiku, and some words of the inscription are legible, such as moon and cool, which are typical words used in haiku.[38]

Even more curiously, there is yet another gravestone near the group of three whale graves. The five-character inscription for this monument appears to be written in an archaic style called "*Awa-moji*." It appears to read "*Kujira-tsuka*" (Whale Grave), but this is not established.[39]

Meanwhile, based on the legend of the white whale associated with Osaki Shrine, the local fishermen believed that whales would save their lives in a shipwreck if they prayed hard enough to the Ebisu God. In fact, a document at the shrine records that when Ōsukaya Yasushirō, aboard the Toyoyoshi-maru bound for Edo (Tokyo), was hit by a storm in 1800, a white whale guided the ship toward the shore and saved the lives of the crewmen. Considering that whales were messengers of the gods, the parishioners of Osaki Shrine did not catch whales or eat them. Also, they have conducted memorial services for whales at the shrine on January 7 or 15 (depending on the source) every year to this day, and offered sacred sake, salt, and sardines to the souls of whales.[40]

Whale Grave at Osaki Shrine, Osaki

In addition, there is another whale grave in the southern part of Kesen'numa, west of the Karakuwa peninsula. It stands in the compounds of Osaki Shrine (or Osaki Myōjin) in Osaki. (This is a different shrine from the one in Karakuwa. The "O" in Osaki is a different Chinese character). This whale grave was created in October 1875. Local history records that fishermen caught a stray whale and sold it to other communities. Using profits from the sale, the community repaired the stone steps of the local Hachiman Shrine and also bought rice for poor families. In appreciation, the community made a grave for the whale at Osaki Shrine.[41]

Memorial Stupa for Whales in Minami-Sanriku-chō, Motoyoshi

Going south from Kesen'numa, there is a monument for whales in Minami-Sanriku-chō, Motoyoshi. This area was totally destroyed by the 2011 Earthquake/Tsunami and became a focus of international media coverage. As of March 2016, local residents are still trying hard to recover from the damage. The monument for whales was erected two hundred and seventy years earlier. The inscription reads:

> "*Kujira kuyō-tō*" (Memorial Stupa for Whales), erected in Enkyō 3 (February 1746–February 1747), contributor, Shōichi of this bay village[42]

The "stupa" refers to a memorial cenotaph, which is usually made of stone, and is erected in the compounds of Buddhist temples, or, in the case of whales, at the shore where a carcass was found. However, the details of the provenance of this monument are undocumented.[43]

Whale Temple in Ayukawa, Ishinomaki

Going down the long strip of the Sanriku Fukkō (Reconstruction) National Park to its southern tip one arrives at the Oshika peninsula, Miyagi prefecture. Oshika-chō was incorporated into Ishinomaki, the second largest city in Miyagi prefecture, in April 2005. Many parts of the city of Ishinomaki, including Oshika-chō, were almost totally destroyed by the 2011 Earthquake/Tsunami. Oshika-chō is located at the eastern tip of Oshika peninsula, and a main community of this mountainous township is the district of Ayukawa at the shore. Also called Ayukawa-hama, it is a very small fishing port. One-and-a-half years after the 2011 Earthquake/Tsunami, a whaling factory ship that had been swept in by the tsunami and run aground on the shore of Ayukawa was left as it was. Only two small whaleships (that looked too small to be whaleships) were moored at the dock.[44]

On the middle of a hill at the shore of Ayukawa stands a whale temple called Shingon School Buddhism Nyoirin-zan Kan'non-ji (Temple of the Goddess of Mercy, "Kan'non Temple" hereafter). Kan'non Temple houses a grand *kujira ihai* (a memorial tablet for whales). "*Ihai*" refers to a lacquered tablet on which a posthumous "afterlife" Buddhist name of the deceased is inscribed, and therefore represents the soul of the deceased. In the case of whales, Kan'non Temple made a collective memorial tablet for whales caught in one season, in lieu of conferring an afterlife name on each whale. Incumbent head priest Kawamura Shigeyoshi states that there used to be many memorial tablets for whales at the temple, but they have been lost in time and now only one remains (a study in 1988 records that there were three at that time). Strangely, although whale grave specialist Yoshihara Tomokichi visited Kan'non Temple, he does not mention them in his study. The inscription on the only existing one reads:

"*Geirei gyorei kuyō-tō*" (Memorial Stupa for the Souls of Whales)[45]

It is about a size of 16 inches in height, whereas an ordinary memorial tablet for a person is usually about six to eight inches. The exceptional size suggests that the local whaling community made a sizable donation to the temple in commissioning it (see Photograph 3.3).[46]

The memorial tablet for people is usually placed at the center of the altar at home, and family members of the deceased offer prayers in front of the tablet every day, whereas the memorial tablet for whales is placed on the altar of Kan'non Temple because whales have no kin. Kan'non Temple houses two altars—one for Buddhist deities and the other for the souls of the dead (*shōryō no ma*). The memorial tablet for whales is placed on the latter, along with memorial tablets for the dead who have no living relatives, as well as collective memorial tablets for the fishermen or whalers who were lost at sea. The head priest offers daily prayers in front of the two altars. Buddhist temples usually conduct a special funeral service called *segaki* (*lit.*, "feeding the hungry ghosts") for those who do not have living kin.[47]

This is based on the belief that since no one is taking care of their afterlives, their souls are tormented and wandering in the world of *gaki* (hungry ghosts) among the Three Worlds of Karmic Reincarnation and that offering this ritual and alms (such as rice and water) will pacify their souls so that they can leave their world and reincarnate as higher beings. Kan'non Temple gives whales an *umi* (sea)-*segaki* because they died at sea and have no kin to take care of their afterlives. The ritual is conducted during the Bon Festival (the week of August 15) when the souls of ancestors and other spirits are believed to return to this world for a week. For this annual ceremony, the town folk carry the memorial tablets for whales and for those who were lost at sea down to the beach and place them on a temporary altar (*segaki-dana*) at the shore. Then, with the attendance of the local fishermen and residents, the head priest performs the *umi-segaki*

Photograph 3.3 Grand memorial tablet for the souls of whales (first on left, back row), Kan'non temple, Ayukawa, Miyagi prefecture

for whales, and sends off their souls to the sea beyond which the souls of the dead are believed to live.[48]

These memorial rites for whales indicate that the local fishermen-whalers treated whales as if they were human beings and made sizable financial contributions to the temple for each service. This in turn reflects the depth of remorse and sorrow on the part of the local fishermen-whalers. The *umi-segaki* was cancelled in 2011 and 2012, due to the destruction of the town by the 2011 Earthquake/Tsunami.[49]

Cenotaphs for One Thousand Whales

In addition, two stone monuments for whales stand in the compounds of Kan'non Temple. One was erected in 1932 by the Isana Company ("Isana" means "brave fish" referring to whales). The inscription for the memorial stone reads:

> "*Sangai banrei-tō*" ("Memorial Stupa for All the Souls in the Three Worlds of Karmic Reincarnation")[50]

This Buddhist phrase refers to the prayer not only for the souls of human beings but also for those of all the living creatures in the world, and in this case referring to whales. Another cenotaph was erected in 1933 by the Ayukawa Whaling Company. The inscription for this monument reads:

> "*Sentō kujira-rei kuyō-tō*" (Memorial Stupa for the Souls of One Thousand Whales)[51]

On the premises, there are also several respectable monuments for those who were lost at sea, as well as large stone statues of *Jizō-bosatsu* (Guardian Deity of Motherhood and Unborn Children).[52]

Whale Shrines in Ayukawa

Meanwhile, three Shinto shrines constituted an integral part of the daily lives of the fishing and whaling community of Ayukawa: Kumano Shrine, Yamadōri-Inari Shrine, and Koganeyama Shrine. Among them, Kumano Shrine is the tutelary shrine of the community and the head priest performs rituals in dedication to the gods of the sea twice a year: the summer and autumn festivals in June and September in the lunar calendar. The head priest of Kumano Shrine is actually descended from a long line of monks at a

Buddhist temple, Shigon Buddhism Mt. Kinka Taikin Temple, on a small island east of Ayukawa which honored Benzaiten (a Japanese Buddhist goddess originated from the Hindu goddess Saraswati, a river deity). The entire island of Mt. Kinka was originally a sacred place where indigenous Japanese gods and goddesses were believed to live and was revered as one of the Three Sacred Sites in the Northeastern Region.[53]

Then, in the mid-twelfth century, Taikin Temple was founded on Mt. Kinka due to the syncretism of Shintoism and Buddhism. Local fishermen considered the whole island as the guardian god for fisheries and safety at sea, and made a pilgrimage to Taikin Temple each time before sailing. However, as the Meiji government enforced the policy of Separation of Buddhism from Shintoism in 1869 and the *haibutsu kishaku* ("abolish Buddhism, destroy Buddha") movement rampaged throughout Japan (see Chap. 2), Taikin Temple was converted into a Shinto shrine and became Koganeyama Shrine. Because of this history, whalers in Ayukawa have also worshiped at Koganeyama Shrine to this day.[54]

Whale Festival and Whale Museum in Ayukawa

More recently, in 1953, the Ayukawa community started an annual whale festival in order to express its appreciation of the catch of the season and to preserve its tradition of whaling as a mode of subsistence. In early August, male residents wearing happi coats carry a palanquin (*mikoshi*) in the shape of a whale and parade through the streets down to the shore. Thus, the whale has become firmly assimilated into the religious and festive lives of this small fishing community, based on a rich spiritual culture of communion with nature. (Several whaling communities in other regions perform similar whale festivals.)[55]

Today, at the shore of Ayukawa stands the Oshika Whale Museum. The building of the museum was partially destroyed (the entrance door was broken) by the 2011 Earthquake/Tsunami, but most of the exhibits appeared to be intact. A local fisherman with the signs on his face of years working on the ocean showed this author around the museum. Among the exhibits, there were several fetuses of whales. This suggests that the local fishermen had a special feeling about the fetuses of pregnant whales they caught, and preserved them. A broken signboard that had fallen on the ground there suggests that this museum is affiliated with the Japanese-government-supported Institute of Cetacean Research in Tokyo, which administers Japanese whaling operations.[56]

Modern Whaling in Ayukawa

True to the signboard, Ayukawa is one of the few fishing communities in Japan currently engaged in legal small-type coastal whaling (STCW) in the near Pacific Ocean, as opposed to the pelagic (open-sea, long-distance) whaling in the Antarctic Ocean banned by the International Whaling Commission. STCW has been conducted on the Sanriku Coast in Miyagi prefecture (with Ayukawa as its base) in the spring and on the coast of Hokkaidō (with Kushiro as its base) in the fall. Although there were no international regulations for this whaling, the Japanese government in 2014 decided to reduce its self-imposed quota for "scientific research" whaling on the Sanriku Coast to 51 from 60. Thus, in April 2014, the small whaleships (that this author had seen moored at the dock in 2012) sailed off for the stated purpose of conducting a survey of the diet of sperm whales—kind and volume—and its impact on fisheries.[57]

Stone Dedicated to the God of Whales

Sailing south from Ayukawa (the island used to be part of Oshika-chō, but was incorporated into Ishinomaki in April 2005) to a small island called Ajishima, there is a stone dedicated to the God of Whales. The local people call this monument "*Kujirappone*" (Bones of Whales). Therefore, it appears that fishermen had buried whale bones there in dedication to the Grand God of Whales. The monument is made of stone with a size of about two feet in height and 1.3 feet in width. The inscription reads:

> "*Tatematsuri kou tasuke kujira-dai-myōjin*" (Respectfully dedicated to and soliciting protection from the Grand God of Whales).[58]

* * *

Whale Monuments on the Sea of Japan

Leaving the Pacific Ocean coast and going west to the Sea of Japan coast, no fishing communities on the western side of the Northeastern Region—corresponding to current Akita and Yamagata prefectures—engaged in classical whaling, and there are few records of monuments for whales in the region, with one exception.

"Whale School" in Akita Prefecture

On the Oga peninsula protruding into the Sea of Japan, northwest of Akita, the capital of Akita prefecture, there is a "Whale School" and a memorial stone for whales. In 1889, more than a hundred whales washed ashore at the fishing village of Funakawa. Using part of the profits from the sale of the whales, the local community constructed a public elementary school. Local residents referred to the school as "*Kujira gakkō*" (Whale School), which has become the Funakawa Daiichi ("first") Elementary School today. In order to show sympathy and appreciation, the community erected a memorial stone for the whales. The monument currently stands on the hill behind the Oga City Office in Izumidai, Funakawa-Minato.[59]

A century later in 1992, the Funakawa community erected a tall memorial stone for hatahata (Japanese sandfish). The area is famous for the catch of hatahata and many local shrines worship this fish as a god, because the fish sustained the livelihood of local residents. (Curiously, the Chinese character for hatahata is made of a left part denoting "fish" and a right part denoting either "god" or "thunder.") The Funakawa community erected a respectable monument for the fish in 1992 when the continental shelf was turned into reclaimed land, because the seaweed bed on the continental shelf was the spawning ground for hatahata.[60]

These monuments in Funakawa indicate appreciation and benevolence for whales and fish on the part of the fishing communities on the Oga peninsula. This in turn suggests that Japanese fishermen on both the Pacific Ocean coast and the Sea of Japan coast shared the same sympathy toward fish and other marine life because they both lived by the principle of coexistence with nature, endorsing the Japanese culture of mourning the deaths of whales.

Notes

1. Natori Takemitsu, *Funka-wan Ainu no hogei* (Ainu Whaling in Funka Bay), Sapporo: Hoppō-bunkasha, 1945; Masami Iwasaki, "Ainu-minzoku kujira-riyō-bunka no sokuseki o tadoru" (Tracing Whale Utilization Culture of Ainu People), *Hokugakuen-daigaku jinbun-ronshū*, No. 21, March 2002, 111–115.
2. Iwasaki, 111–115, 121–128.

3. Ibid., 121–128; Kumano Taiji-ura hogei-shi hensan-iinkai, ed., *Kumano no Taiji: Kujira ni idomu machi* (Taiji, Kumano: Town That Challenged Whales), Tokyo: Heibonsha, 1965, 183–184.
4. Ibid. (both).
5. Iwasaki, 132–142.
6. Komatsu Masayuki, *Yutakana Tokyo-wan: Yomigaere Edo-mae no umi to shoku-bunka* (Rich Tokyo Bay: Restore the Sea and Food Culture of Edo), Tokyo: Yūzankaku, 2007, 91–95.
7. "Hakodate shinkei" (A Realistic View of Hakodate), *Nihon hogei zu-reki* (Picture Calendar of Japanese [Classical] Whaling), Tokyo: Institute of Cetacean Research and Japan Whaling Association, 2005; Boreal Institute for Northern Studies, ed., *Small-Type Coastal Whaling in Japan: Report of an International Workshop*, Edmonton, Alberta: Boreal Institute for Northern Studies, the University of Alberta, 1988, 12–13.
8. Boreal Institute for Northern Studies, 63–64; Masami Iwasaki, "The Cultural Significance of Whaling in a Whaling Community of Abashiri," M.A. thesis, Department of Anthropology, University of Alberta, Edmonton, 1988, 66–72.
9. Boreal Institute for Northern Studies, 62–64.
10. Ibid.
11. Ibid.
12. "Kujira-zoku kuyō-tō" (Cenotaph for Genus of Whales), http://www.kanchu-h.ed.jp/zennichi/main_kikaku/hakodate/newslope/kujira.htm, accessed October 12, 2015.
13. Ibid.
14. Ibid.; "Kujira-zoku kuyō-tō" (Cenotaph for Genus of Whales), July 4, 2004, http://www.asahi-net.or.jp/~rn2h-dimr/ohaka2/12kujira/11hokaido/hakodate.html
15. "Green Reconstruction: Creating a New National Park," http://www.env.go.jp/jishin/park-sanriku/images/sanriku_fukkou_project_eng.pdf, downloaded March 11, 2016.
16. Yoshihara Tomokichi, "Kujira no haka" (Graves of Whales), in Tanigawa Ken'ichi, ed., *Nihon minzoku bunka shiryō shūsei* (Collection of Documents on Japanese Folk Culture), Vol. 18, Tokyo: San'ichi-shobō, 1997, 423.
17. Ibid.
18. Ibid.
19. Ibid., 421; "*Todo kuyō-hi*" ("Memorial Stone for Steller Sea Lions"), July 25, 2011, http://www.jpcoast.com/entry/8595.html
20. Ibid. (both).
21. "*Sake kuyō*" ("Memorial for Salmon"), June, 2011, http://www.jpcoast.com/entry/8594.html

22. "Rikuzentakata's 'Miracle Pine Tree' Now Restored," March 7, 2013, http://japandailypress.com/rikuzentakatas-miracle-pine-tree-now-restored-0724729/
23. Yoshihara, 421–422.
24. Ibid.
25. "Kodate kan'non-dō" ("Little Prayer House for the Goddess of Mercy"), *Tōkai-shimpō*, November 5, 2009; "Gyoran kan'non" (Goddess of Mercy for Fish), May 19, 2012, http://seiganjiblog.blog104.fc2.com/blog-entry-116.html
26. Ibid. (both).
27. Ibid. (both).
28. "Kodate kan'non-dō"; emails to author, July 3, July 9, and July 17, 2012.
29. Emails to author, July 3, July 9, and July 17, 2012.
30. "Kodate kan'non-dō."
31. "Gyogyō no machi de sakana-kuyō" (Memorial Service for Fish at Fishing Town), *Tōkai Shimpō*, September 24, 2010.
32. Ibid.
33. "Nihon no kujira bunka: Kujira to shinkō" (Japanese Whale Culture: Whales and Beliefs), http://www.geocities.co.jp/NatureLand-Sky/3011/kujira-to-sinkou.html, accessed November 23, 2015.
34. Yoshihara, 422.
35. "Osaki-jinja" (Osaki Shrine), October 19, 2013, http://www.genbu.net/data/mutu/osaki_title.htm
36. Yoshihara, 422–423.
37. Ibid.
38. Ibid.
39. "Karakuwa-hantō Osaki no 'Kujira-zuka' to kodaimoji no tsuka" ("Whale Grave in Osaki at Karakuwa Peninsula and Grave with Inscription in Ancient Letters"), January 20, 2008, http://blog.goo.ne.jp/tako_888k/e/70e5e29a67125b216cc51334f20fc195
40. Ibid.; Yoshihara, 422–423. Yoshihara states that the memorial service is conducted on January 15, while the blog states January 7.
41. Yoshihara, 423.
42. "Kujira bunka: Kujira o tomuratta kujira-baka · kujira-zuka nado" (Whale Culture: Whale Graves and Monuments to Mourn the Deaths of Whales), http://www.geocities.co.jp/NatureLand-Sky/3011/kujirahaka.html, accessed November 14, 2015; "Kujira kuyō-tō" (Memorial Stupa for Whales), http://www.town.minamisanriku.miyagi.jp/museum/life/article.php?p=411, accessed December 4, 2015.
43. Ibid. (both).
44. Ayukawa-hama Town Hall, email to the author, June 25, 2012; author's own observations.

45. Kawamura Shigeyoshi, author interview, August 7, 2010; Boreal Institute for Northern Studies, 57–58.
46. Ibid. (both).
47. Ibid. (both).
48. Ibid. (both).
49. Ibid. (both).
50. Yoshihara, 423.
51. Ibid.
52. Author's own observations.
53. Kawamura; Boreal Institute for Northern Studies, 60–61; "Kinka-zan Koganeyama-jinja" (Mt. Kinka Koganeyama Shrine), http://kinkasan.jp/about/yuisyo, accessed March 3, 2016.
54. Ibid. (all three).
55. Boreal Institute for Northern Studies, 61–65.
56. Author's own observations.
57. "Sanriku-engan no chōsa-hogei e shukkō: hogei-sū o herashite jisshi" (Research Whaleships off Sanriku Coast Departed: Implementing Reduced Quota), *Asahi Shimbun*, April 26, 2014.
58. Yoshihara, 423–424; "Kujira bunka: Kujira o tomuratta kujira-baka · kujira-zuka nado."
59. "Oga no Kuyō-hi" (Memorial Stones in Oga), http://www.city.oga.akita.jp/index.cfm/14,2682,c,html/2682/20120528-132,439.pdf, accessed December 21, 2015.
60. Ibid.

CHAPTER 4

Whale Shrines and Temples in the Kantō Region

There was a notable classical whaling base in the Kantō Region (central-eastern Honshū) on the Pacific Ocean coast—encompassing such prefectures as Chiba, Tokyo, and Kanagawa. The Daigo clan organized a whaling guild on the southern Bōsō peninsula and engaged in coastal whaling, while local fishermen caught strayed or stranded whales that were migrating offshore. Consequently, there are a number of distinctive whale graves and monuments in the region.

History of Classical Whaling in Chiba Prefecture

Local history records that in 1612 the second lord of the state of Awa-Tateyama in the southern Bōsō peninsula (current southern Chiba prefecture), Satomi Tadayoshi (1594–July 1622), donated a whale skin to Ise Shrine (in current Mie prefecture) as the first catch and produce of the season. This suggests that whales were still considered sacred in the seventeenth century. In 1655, Daigo Shinbee in Hota village in Awa-Katsuyama (current Kyonan-machi, Awa county) founded a local whaling guild called the Daigo-gumi, which continued its operations for ten (or eleven) generations until the Meiji period (1868–1912). This was the sole classical whaling guild that existed in the Kantō Region. Over the course of time, the whaling base on the Bōsō peninsula was moved from Katsuyama to Shirahama, Chōshi, Chikura, and to Wada.[1]

Today, all the classical whaling bases on the *Bōsō peninsula* have ceased to exist, with the exception of Wada-machi (renamed Wada-chō, Minami-Bōsō

M. Itoh, *The Japanese Culture of Mourning Whales*,
https://doi.org/10.1007/978-981-10-6671-9_4

in March 2006), which engages in *small-type coastal whaling* (STCW), catching Baird's beaked whales. This smaller species of whale is not banned by the International Whaling Commission's (IWC) moratorium on commercial whaling, but the Japanese government enforces its own regulations and quotas, and the Wada Whaling Union operates in compliance with them.[2]

Whale Grave at Kawaguchi Shrine in Chōshi

Traveling the Bōsō peninsula from northeast to northwest, a local document in 1953 records that there was a whale grave in Kawaguchi Shrine in Chōshi, Chiba prefecture, at the northeastern edge of the peninsula. The local fishermen regarded whales as the Ebisu God (the God of Fisheries) and buried the head of a whale at the shrine as a memorial to the whale and in dedication to the Ebisu God. The whale grave weathered away in time, and currently there is no vestige of the grave at the compound. Meanwhile, the local fishermen also created a grave for sea turtles, and a gravestone for sea turtles currently exists at the shrine.[3]

Whale Grave in Chikura-machi, Minami-Bōsō

Toward the south of the Bōsō peninsula, one finds a respectable whale grave in the compound of Shingon School Buddhism Chōshō Temple located in the Senda district in Chikura-machi (incorporated in a newly created city, Minami-Bōsō, in March 2006). A whale washed ashore in Senda port in September 1896. The local residents considered the whale to be a gift from the Ebisu God, gratefully shared the meat, and planned to make a grave for it. They originally wanted to bury all the remains and consulted with the head priest of Chōshō Temple. But the remains were too large to bury in the compound of the temple, and they decided to bury only the whale's heart.[4]

The residents also wanted to make a special gravestone for the whale and decided to use an old gravestone on the premises of the temple, which was abandoned and no living relatives took care of it. The gravestone was made in the style of "*hōkyōin-tō*": a version of the Buddhist stupa with five basic tiers of curved stones in different shapes called *gorin-tō* (a five-tiered stupa), but having a more elaborate design and structure than a *gorin-tō*. The inscription for this gravestone, "*Kujira-zuka*" (Whale Grave), was made at the bottom tier called the "base of the stupa." It is still legible today. Currently, the gravestone stands by the stone steps leading to the

Kan'non-dō, where a statue of the Goddess of Mercy is housed. Chōshō Temple also has an impressive, grand bronze hanging bell, guarded by beautiful and yet solemn looking statues of Six *Jizō-bosatsu* (Guardian Deities of Motherhood and Unborn Children).[5]

Whale Graves in Shirahama-machi, Minami-Bōsō

At the southern tip of the Bōsō peninsula stand two whale gravestones in Otohama, Shirahama-machi (incorporated into a newly created city, Minami-Bōsō, in March 2006, along with Chikura-machi and other townships). There is somewhat conflicting information on these gravestones. According to the study in 1982 by Yoshihara Tomokichi, a local fisheries specialist by the name of Fuchibe Tomoyuki found the skull of a whale and the body of a fetus in the ground around 1916. He reburied them and created graves for them. The two gravestones stand on a large foundation made of blocks of stone, but there is no inscription on the gravestones. A *torii* (sacred entrance gate at the shrine) used to stand in front of the graves, but it weathered and decayed.[6]

In recent years, a new stone signboard (undated, but sometime after 1988) was installed on the foundation of the two gravestones. Its inscription gives a more accurate provenance for the graves. It reads: "*Kujira-zuka*" (Whale Grave).

Coastal whaling in Japan has a history of over two hundred years, and the Otohama district of Shirahama had also engaged in whaling. The site of the condominium building behind this monument used to be the location of the Tōkai Gyōgyō (Fisheries) Company. The two gravestones, as well as the skull bones of the mother whale and the body of its fetus, were found underground at that site at the time of construction of the condominium building. The gravestones were then transferred to the front of the lot, while the skull of the mother whale and the body of its fetus have been preserved at the Shirahama town government office. Local history has it that the gravestones were created around 1871 as a memorial to a mother whale and its fetus. Since then, each time the company set out for whaling, the crew members first gathered in front of this monument and prayed for the souls of the whales and for safe sailing.[7]

Cube Memorial Stupa for Whales in Shirahama

In addition, there is a "*kaku-tōba*" (a memorial stupa in the shape of a cube) for whales in Shirahama. The owner of the local fisheries company, the Tōkai Gyōgyō Company, Aoki Orito, erected this monument in 1947 for the souls of whales. The inscription for this monument had weathered away and became illegible by 1982, but according to a pioneering study of Shindō Naosaku (1970), the inscription read:

> Cube Memorial Stupa…For the sake of stopping the suffering of the dead who have no living relatives to take care of their afterlives, as with the whales, so that their souls may attain the Three Enlightenments of Buddha. Donated ¥100,000, June 16, 1947[8]

Shirahama Oceanic Art Museum

In addition, the Shirahama Oceanic Art Museum stands near the Noshimasaki Lighthouse, at the southernmost tip of the Bōsō peninsula. This private museum houses various artifacts relating to fisheries, as well as many paintings, including a picture scroll of classical whaling. It also has the originals of famous portraits of US commodore Matthew C. Perry (April 1794–March 1858) and of Russian admiral Yevfimy Putyartin (November 1803–October 1883). On the premises lies "*Kujira-iwa*" (Whale Rock), a huge rock that looks as if a life-size whale is lying there. This is actually the grave of a local medical doctor and haiku poet, Inoue Suginaga (August 1770–February 1828), and was made in 1833 by his disciples. This is a reminder that the site of the art museum used to be that of a temple. A large gravestone marker, "*Suginaga-fun*" (Suginaga Grave), is erected on top of the Whale Rock.[9]

Benzaiten Shrine in Katsuyama, Kyonan-machi

Moving on to the western side of the Bōsō peninsula, a grave for a whale fetus lies on the shore of Awa-Katsuyama-machi, Awa county. (The town was renamed Kyonan-machi in March 1959, but the old town name is retained as Awa-Katsuyama JR East Japan station today.) Awa-Katsuyama was the whaling base of the Daigo-gumi, the whaling guild founded in 1655 by Daigo Shinbee. It operated only during the summer as whales migrated off the Bōsō peninsula only during that season. The Daigo-gumi continued to employ the manual harpoon method, even after the net method was invented by Taiji Kakuemon in 1675 (see Chap. 6). The net method was unsuitable to catch Baird's beaked whales because they dove deep to feed on deep-sea squid. Meanwhile, the Daigo clan founded the clan's own temple, Daigo-san Myōten Temple, where the "Daigo Shinbee(s)" of many generations are buried.[10]

The Daigo clan also founded Benzaiten Shrine (where Benzaiten, one of the Seven Gods of Fortune in the Japanese syncretism of Shintoism and Buddhism, is enshrined, see Chap. 2) in Itaigayatsu in the current Katsuyama district of Kyonan-machi. There, the Daigo clan erected a memorial stone, "*Kujira Benzaiten-hi*" (Monument of Benzaiten for Whales), in dedication to Benzaiten and the Ebisu God.[11]

Whale Graves in Katsuyama, Kyonan-machi

Astonishingly, more than 120 whale graves (*kujira-zuka*) existed in Itaigayatsu Benzaiten Shrine (also called Itsukushima Shrine). Throughout Japan, many shrines that worship Benzaiten are called Itsukushima Shrine, including the famous one on Miyajima in Hiroshima prefecture. Many of the whale graves in Awa-Katsuyama have been destroyed by rockfalls and other natural disasters. At present, 52 whale graves stand in front of the cliff next to the *hokora* (a small shrine) for Benzaiten and the Ebisu God that Daigo Shinbee VII, Jōkō, donated in 1838. A signpost, *Kujira-zuka* (Whale Graves), stands at the entrance. The graves are in the style of "*kasa-tōba*" (a memorial stupa in the shape of a cube with a hat-like roof on top). At 2.7 feet in height (there some shorter ones) × 1.0 foot in width × 0.7 feet in depth, the gravestones are relatively small.[12]

Beginning around 1612, the graves were created by local fishermen—members of a guild called the "Deba-gumi" ("*deba*" refers to a sharp bladed knife)—who had engaged in seasonal whaling. They made a whale grave after each season and donated it to the shrine in order to express their appreciation to the Ebisu God and also pray for the souls of the whales they had caught. The size of the gravestones differed each season. They made a larger gravestone for years with a big catch and a smaller one for less successful years.[13]

These gravestones at Itaigayatsu Benzaiten Shrine are smaller compared to those found in the western regions of Japan, such as in the Saikai Classical Whaling Base in Kyūshū. This is probably because the Daigo-gumi was not directly involved in the actual catching and disassembling of whales, but only took part in the processing aspects of whaling—taking whale oil and making bone powder. In contrast, the whaling guilds in the Saikai Classical Whaling Base took charge of all aspects of whaling, from catching and disassembling to processing. Therefore, instead of the Daigo clan, the local fishermen of the

Deba-gumi made the graves for whales because they were directly involved in killing them. Although the gravestones are smaller than those in other regions, corresponding to the financial situation of the Deba-gumi, it is quite a sight to see scores of whale gravestones standing next to each other by the cliff, being covered with ferns and other greens. In turn, the Daigo clan did its part by founding Benzaiten Shrine itself and donating monuments to the shrine.[14]

Ema at Kachiyama Shrine

In 1836, Daigo Shinbee VII, Jōkō, donated a huge *ema* ("picture horse") to Kachiyama Shrine. The design of this ema is unique. The picture on this wooden plaque is a famous scene from *Kojiki* (the oldest history record of Japan, complied in 712) about the mighty storm god of the sea, Susan-ō-no-mikoto (third son of Izanagi and Izanami, to whom the creation of Japan was attributed) challenging the eight-headed dragon, Yamata-no-orichi, in the upper reaches of the Hi River in the state of Izumo (current Shimane prefecture), in order to save the princess Kushinada. Daigo donated the giant *ema* because Tenpō 7 (1836) was a peak year of the "Great Famine of Tenpō," which lasted from 1833 to 1837/1839.[15]

While countless Japanese suffered all over the country, residents in the Katsuyama bay village were saved by a good catch of whales that year. Daigo therefore donated the giant *ema* in order to express appreciation to the Ebisu God and to pray for the souls of the whales. Today, this giant *ema* is preserved at the Hishikawa Moronobu Memorial Museum in Kyonan-machi, an art museum named for Hishikawa Moronobu (1618–July 1694), the founder of the art genre, *ukiyo-e* (*lit.*, "pictures of the floating world"), who was born in this locale.[16]

Reflecting the tradition of classical whaling in Awa-Katsuyama, Kyonan township has celebrated the whale festival in which local men parade the streets carrying a palanquin (*mikoshi*) in the shape of a whale and take it down to the shore. The scene was featured in the 2014 Japanese film, "*Niji no misaki no kissaten*" (*lit.*, "café at cape rainbow"; English title, Cape Nostalgia).[17]

Whale Shrine in Urayasu

Moving further north along the Boso peninsula coast all the way up to Urayasu, which is adjacent to Tokyo across the Edo River, there is an interesting whale shrine at Inari Shrine in the town of Tōdai-jima (*lit.*, "island of this generation"). As the name suggests, this area used to be in the sea and Urayasu was a fishing community, but the area has become reclaimed land and Urayasu has become a large city. (Today it is a popular tourist destination as the location for the Tokyo Disney Resort). At Inari Shrine, there is a separate shrine (from the main shrine) specifically made for a single whale called "*Ō-kujira no o-yashiro*" (Shrine for Big Whale).[18]

This whale shrine houses four memorial stones for the whale in the style of *kasa-tōba* (a memorial stupa in the shape of a cube with a hat-like roof on top). Two of them are without inscriptions, while one has the inscription, "*Ō-kujira*" (Big Whale), and the other has an inscription in vermillion paint "*Sanuki-koku Kotohira-gū*" (Kotohira Shrine, Sanuki State). Kotohira Shrine, in current Kagawa prefecture on Shikoku Island, is the supreme head shrine of all the Kotohira Shrines in Japan, which enshrine a guardian god of maritime traffic. The Japanese refer to it as "Konpira-sama" and more than three million people, with an increasing number of foreign visitors, annually make a pilgrimage to the shrine despite its remote location.[19]

An auxiliary monument for the whale shrine describes its provenance; it reads:

> In 1875 two local fishermen, Takanashi Genpachi and Nishiwaki Seikichi, caught a large whale in the bay of Urayasu and sold it. This whale was towed to Fukagawa, in Edo (current Tokyo), and was shown to the Edoites and caused a sensation. The two fishermen were treated as heroes so that they slacked off from their work afterwards. In order to cool down from the whale fever and focus their minds back on their work, they took advice from an elder and donated the whale shrine and memorial stones for the whale to the local Inari Shrine. Further, in 1925, a wooden plaque was donated to this whale shrine, with the inscription, "*Hōnō Ryūgū*" (Donated in dedication to Ryūgū, the undersea palace of Dragon God of the Sea).[20]

Whale Grave in Urayasu

It is recorded that around 1892 a whale ran aground on the shore of the fishing town Urayasu. A show proprietor bought it in order to display it to the public in Asakusa, in downtown Tokyo. There was a famous, eclectic amusement park in Asakusa, called Hanayashiki, with a botanical garden, a theater for vaudeville performances, and a zoo (the amusement park still exists today). However, the whale was too heavy to be transported to Asakusa, and its body eventually decayed. The showman commissioned the local Buddhist temple, Kezō-in, to bury the whale on its premises, despite the fact that this kind of service at a temple required a considerable financial contribution (euphemism for a fee) and also involved a funeral service. The temple also erected a memorial stupa at the whale grave. Unfortunately, the memorial stupa weathered away in time and does not exist any longer.[21]

* * *

Whale Graves and Monuments in Tokyo Prefecture

Interestingly, there were at least two whale graves in Tokyo (formerly Edo), as some whales had drifted deep into Edo Bay (current Tokyo Bay). In fact, the local history records that stray whales reached the shores of Edo Bay at intervals of about thirty-years:

1. two whales drifted onto the shore at Gyōtoku-Kōya village (in current south of Ichikawa, Chiba prefecture) in March 1734. One of them, a 32-foot-long whale, was publicly displayed in Ryōgoku in eastern Edo and Edoites flocked to see it;
2. a stray whale drifted onto the shore of Kanagawa (current Yokohama, Kanagawa prefecture) in May 1772 and people gathered to see it, but they became sick due to the strong odor;
3. a 55-foot-long, 6.8-foot-tall stray whale was stranded ashore in Shinagawa and was caught in Ten'nōzu in June 1798;
4. two 39-foot-long whales drifted off Fukagawa in March 1820; and
5. an about 18-foot-long whale washed ashore of Ōi-Ohayashi-machi in Samezu Bay in May 1851. Few people went to see the whale because it smelled badly.[22]

"Whale Incident of Kansei"

The huge 55-foot-long whale that was caught in Ten'nōzu, Shinagawa, in June 1798, caused a sensation, so that the event was referred to as the "Whale Incident of Kansei," after the Kansei period (February 1789–March 1801). Ten'nōzu was like a small island surrounded by the sea at that time, as many bay areas of Edo were, but all of the areas have become part of reclaimed land now. In 1798 the news of catching a gigantic whale reached Eleventh Shogun Tokugawa Ienari (November 1777–February 1841) in Edo Castle. Subsequently, the whale was towed by boats to the Hama Detached Palace (current Hama Detached Palace Park, adjacent to the Tsukiji Fish Market) for a private viewing by the incumbent shogun.[23]

Then, the whale was displayed for a public viewing off Shinagawa. The Edoites flocked to see the whale, and the fee for a ferry ride to Shinagawa skyrocketed to 100 mon, a 500-fold increase. Even a *kyōka* (a poem infused with parody) was written (author unknown), attesting to the popularity of the whale:

> A *semi-kujira* anchored off Shinagawa
> Every one is flying to see it, with the buzzing sound, *min min*[24]

The word "*semi-kujira*" (*lit.*, "beautiful back whale") refers to the North Pacific right whale. But the word "*semi*" also refers to a cicada, and there is a species of cicada called the "*min min zemi*" (*zemi* is a phonetic variant of *semi* as with "liaison" in the French language), named after its chirping sound, *min min*. Therefore, "*semi-kujira*" at the beginning of the poem forms a pun with "*min min zemi*" (*semi*) at the end.[25]

The "Whale of Kansei" was one of three animals to have caused a sensation among Edoites during the Edo period, along with the "Elephant of Kyōho"—an Asian elephant pair from Vietnam brought to Nagasaki in 1728; the female soon died, but the male elephant walked all the way to Edo from Nagasaki, and was presented to the Eighth Shogun Tokugawa Yoshimune in 1729—and the "Arabian Camels of Bunsei"—the Dutch traders shipped a dromedary pair from Mecca (in current Saudi Arabia) to Nagasaki in 1821, but the Tokugawa shogunate government refused to buy them, and they were sold to a circus entertainer and toured all over Japan.[26]

Whale Grave at Kagata Shrine

The body of the "Whale of Kansei" was left to decay afterwards, and local fishermen buried some of its bones and erected a gravestone for the whale in the nearby Suzaki-Benten Shrine at the shore of Shinagawa Bay (in current Higashi-Shinagawa, Shinagawa ward). There are famous woodblock prints of various scenes of Suzaki-Benten Shrine by Andō Hiroshige. The shrine was renamed as Kagata Shrine after the surname of a person who started land reclamation in the area. The gravestone the fishermen erected at the shrine is made of aqueous rock in the shape of a pyramid about 2.6 feet high × 5 feet in width. The vermillion red engraving on the gravestone reads: "*Kujira-hi*" (Memorial Stone for Whale).[27]

The rest of the engraving in smaller characters has weathered away and is illegible now, but an auxiliary monument made in 1798 describes the provenance of the grave. The inscription reads:

> Memorial Stone for Whale. Erected by fishermen of Ten'nōzu, Shinagawa Bay, Ebara county, Bushū. Whales are the kings of the fish. They are abundant in the sea in southwestern Japan, but rare in the sea in the northeast. A stray whale swam off Tene'nōzu this summer and the fishermen caught it. They immediately reported it to the government. People heard about it and flocked to see the whale for several days...
>
> A summer whale, drifting into Shinagawa, earned a great fame in Edo June 1798[28]

The haiku poem at the end of the monument was written by Tani Sogai (1717–1809). This whale grave is significant as the only one of its kind that currently exists in Tokyo.[29]

New Memorial Stone for Whale at Kagata Shrine

Realizing the historical significance of the "Whale Incident of Kansei" and the whale grave at Kagata Shrine, the local town community of Higashi-Shinagawa founded the Whale Grave Preservation Society and created a new monument in September 1969. Its inscription, "*Kujira-hi*" (Memorial Stone for Whale), is also written in vermillion, after the original gravestone. Below the title is another inscription written in modern Japanese describing the provenance of the original whale grave. This memorial stone stands on a foundation adorned by the colorful figure of a dragon, from which the water runs down to a carp pond. Today, the original gravestone and the new monument stand next to each other. Also, a small stone with the inscription "*Kujira-zuka*" (Whale Grave) is placed in front of the original gravestone in order to indicate that there was actually a whale grave there.[30]

The preservation society also erected new signboards that describe the provenance of the whale and the whale grave, as well as the history and lifestyle of the Edoites in the Shinagawa bay area at the time. The signboard also explains that the fishing community had to report each whale caught and pay a tax for it to the presiding province, even if they did not belong to any whale guild and did not professionally engage in classical whaling. Catching whales was not exactly "finders keepers" because it did not happen every day and it brought sizable tax revenue to the province. Further, the preservation society created in the playground adjacent to the shrine a statue of a head of a whale and riding equipment for children in the shape of a whale.[31]

Judging from the specifications of the whale recorded by a shogunate government official and the drawing of the whale made at that time, it appears that the whale might actually have been a blue whale, not a North Pacific right whale, as suggested by the *kyōka* poem (shown above). The poet might have portrayed the whale as a *semi-kujira* (North Pacific right whale) for the sake of the pun of *semi* (cicada). The actual drawing of the whale has a dorsal fin and also ventral pleats that extend form the throat to the navel. These are features of blue whales, whereas North Pacific right whales have neither dorsal fins nor ventral pleats.[32]

Whale Grave at Samezu Hachiman Shrine

Local history recorded that there was another whale grave in Edo. This was for an 18-foot-long whale that washed ashore at Samezu Bay of Ōi-Ohayashi-machi (current Higashi-Ōi, Shinagawa ward) in May 1851. Local fishermen buried the body in the compound of Samezu Hachiman Shrine; however, it weathered away in time and no vestige exists today.[33]

* * *

Whale Monument in Kanagawa Prefecture

Whale Grave in Misaki, Miura

South from Tokyo, there is a whale grave in Misaki, Miura, at the tip of the Miura peninsula. This grave was created for a stray whale that was caught near the Jōgashima Lighthouse in March 1834. Local fishermen created a grave for the whale in the local Nishihama Jizō-dō Temple and erected a gravestone with a size of about 2.2 feet long × 1.0 foot wide. The inscription reads: "*Kujira-zuka*" (Whale Grave).[34]

On the side of the gravestone are engraved the names of four financial contributors for making the gravestone, as well as the date. This gravestone is now housed in a structure made with blocks of stone, along with a large stone statue of *Jizō-bosatsu* (Guardian Deity of Motherhood and Unborn Children) in the center and a couple of smaller statues of *Jizō-bosatsu* on the side. The caretakers of the temple have maintained the monuments well, cleaning its and offering fresh flowers, suggesting local people's benevolence toward whales.[35]

Notes

1. "Awa no hogei (koshiki–genzai)" (Whaling in Awa [Classical–Present]), http://www.geocities.co.jp/NatureLand-Sky/3011/bousou-hogei.html, accessed December 4, 2015; Boreal Institute for Northern Studies, ed., *Small-Type Coastal Whaling in Japan: Report of an International Workshop*, Edmonton, Alberta: Boreal Institute for Northern Studies, the University of Alberta, 1988, 26–27.
2. Ibid. (both).
3. Yoshihara Tomokichi, "Kujira no haka" (Graves of Whales), in Tanigawa Ken'ichi, ed., *Nihon minzoku bunka shiryō shūsei* (Collection of Documents on Japanese Folk Culture), Vol. 18, Tokyo: San'ichi-shobō, 1997, 428.
4. Tokyo-wangan nai · gai no kujira-zuka" (Whale Graves Around Tokyo Bay), http://www.geocities.co.jp/NatureLand-Sky/3011/haka-toukyou.html, accessed December 5, 2015.
5. Ibid.; "Chōshō-ji no kujira-zuka (Whale Grave at Chōshō Temple), September 13, 2008, http://blog.goo.ne.jp/yoshi883t/e/505d84c4a6af9916640600b15818c199
6. Yoshihara, 427–428; Boreal Institute for Northern Studies, 54.
7. "Tokyo-wangan nai · gai no kujira-zuka"; Boreal Institute for Northern Studies, 54.
8. Shindō Naosaku, *Seto-naikai shūhen no kujira no kenkyū* (*Study of Whales Surrounding the Inland Sea*), Kobe: Ikuta-ku ishikai, 1970, 141; Yoshihara, 27–428.
9. "Shirahama kaiyō bijutsukan" (Shirahama Oceanic Art Museum), http://homepage2.nifty.com/kaibi/, accessed December 8, 2015.
10. "Awa no hogei (koshiki–genzai)"; Komatsu Masayuki, *Yutakana Tokyo-wan: Yomigaere Edo-mae no umi to shoku-bunka* (Rich Tokyo Bay: Restore the Sea and Food Culture of Edo), Tokyo: Yūzankaku, 2007, 79–83.
11. Ibid. (both).
12. Ibid. (both).
13. Ibid. (both).
14. "Awa no hogei (koshiki–genzai)."
15. Ibid.
16. Ibid.
17. "Chiba Kyonan-machi no kissaten o eiga-ka" (Film about a Café in Kyonan-machi, Chiba), *Asahi Shimbun*, March 7, 21,014.
18. "Tokyo-wangan nai · gai no kujira-zuka."
19. Ibid.

20. Ibid.
21. Yoshihara, 428.
22. Ibid., 424–425 and 428.
23. Ibid., 434–435; "Tokyo-wangan nai · gai no kujira-zuka."
24. Ibid. (both).
25. Ibid. (both).
26. Komatsu, 90; "2011-nen: Sōgō-toshokan kichōsho-ten" (2011 Exhibition of Rare Books at [Tokyo University] General Library), http://www.lib.u-tokyo.ac.jp/tenjikai/tenjikai2011/dobutsu.html, accessed December 6, 2015.
27. Yoshihara, 424–425.
28. Ibid.
29. Ibid.
30. "Kagata-jinja to kujirazuka" (Kagata Shrine and Whale Grave), August 4, 2007, http://www.asahi-net.or.jp/~rn2h-dimr/ohaka2/10gyogyo/13kanto/kagata.html
31. Ibid.
32. "Tokyo-wangan nai · gai no kujira-zuka."
33. Yoshihara, 424–425.
34. Ibid., 427; "Tokyo-wangan nai · gai no kujira-zuka."
35. Ibid. (both).

CHAPTER 5

Whale Graves and Monuments in the Chūbu and Hokuriku Regions

This chapter examines whale graves and monuments in two regions: the Chūbu Region on the Pacific Ocean coast, and the Hokuriku Region on the Sea of Japan coast, which present distinct whaling cultures and traditions from those in the Chūbu Region.

Chūbu Region

Moving west from the Kantō Region to the Chūbu Region (central part of Honshū) on the Pacific Ocean coast, there is no record of whale graves or monuments in Shizuoka and Aichi prefectures. While it is known that fishing villages in Mikawa province and Owari province (both in current Aichi prefecture) engaged in classical whaling on a small scale, no record has been found of whale monuments.

Classical Whalers in Owari Province

In the mid-sixteenth century, there were a number of skilled classical harpooners in the State of Owari (the current western part of Aichi prefecture). However, geographic conditions of Owari were not suited for whaling because Ise Bay, which surrounds the area, is an enclosed bay and was away from whale migration routes. The harpooners in Owari therefore tried to establish their whaling bases in other regions. For instance, a journal written by Miura Jōshin Shigenobu in the late sixteenth century records that during the Bunroku period (January 1593–December 1596) an expert

M. Itoh, *The Japanese Culture of Mourning Whales*,
https://doi.org/10.1007/978-981-10-6671-9_5

harpooner in Owari by the name of Mase Sukebee observed that many whales migrated off the Miura peninsula in the State of Sagami (current Kanagawa prefecture) and began classical whaling there.[1]

In the beginning, Mase Sukebee was blessed with a good catch, and the local fishermen in the neighboring villages learned the harpoon method from him so that the catch reached one hundred to two hundred in one year. However, the catch gradually decreased and had fallen twenty years later to only three or four whales a year. Mase stopped whaling there. Another document records that in subsequent years a fisherman from Owari began whaling in Tango (current Kyoto prefecture) and Tajima (Hyōgo prefecture) on the coast of the Sea of Japan. However, the catch was poor and he could not pay the operating costs. He gave up whaling there and returned home.[2]

In addition, there is a record that in 1600 a classical whaling guild in Owari called the Owari-Senga-gumi had three whaleboats crewed by nineteen crewmen, including a foreman. The guild was based in Morozaki village (current Minami-Chita-chō, Chita country, Aichi prefecture) located at the southernmost tip of the Chita peninsula (today, the Chūbu Centrair International Airport sits in the middle of the peninsula). The Owari-Senga-gumi existed until the end of the Edo period (1603–1868). However, the guild did not have its own established whaling grounds, because their grounds in Ise Bay were unsuitable for whaling. The crewmen therefore sometimes crossed the boundary of the territorial waters of Owari province and sailed to waters off Kishū province (current Mie prefecture).[3]

Nevertheless, Kishū province was lenient with the violation of the crossing of the territorial waters by the Owari-Senga-gumi, because Owari province and Kishū province were "relatives," being respectively ruled by the ninth son (Tokugawa Yoshinao) and the tenth son (Tokugawa Yorinobu) of Tokugawa Ieyasu, the founder of the Tokugawa shogunate government.[4]

In 1606, Wada Chūbee Yorimoto (Wada Kin'emon, I) of Taiji (the current Wakayama prefecture) organized a harpooners' whaling guild, with the assistance of a fisherman from Morozaki, Owari, by the name of Denji (see Chap. 6). In 1618, Yorimoto's son, Wada Kin'emon II, Yoriteru, hired a fisherman from Onoura near Morozaki, Owari, by the name of Yoheiji, who was a skilled harpooner. Wada gave Yoheiji the position of "*hazashi*" (*lit.*, "blade thruster"). This is said to be the origin of the job title name for the diver who jumped naked into the freezing sea, first among the crewmen, and cut into the whale's blowhole in order to tie a fishing net through it (similar to making a cut in the nose of an ox in order to attach a cowbell).[5]

In addition, around 1644, a classical whaler in Owari province, Oike Gizaemon, sailed to Tosa (in current Kōchi prefecture) on Shikoku Island and observed that many whales migrated off Tosa Bay. Oike sent for his family member, Oike Shirōemon, and together they began classical whaling there. Their operations succeeded to a certain degree but did not last long, and they established a new whaling base in Tosa-Saga at the western edge of Tosa Bay (see Chap. 8).[6]

These historical records suggest that the local harpooners of Owari province tried in earnest to establish their whaling bases in other fishing communities far away from their home territory, while the Owari-Senga Whaling Guild crewmen sailed far from their own base to catch whales. The whalers in Owari could only catch whales that drifted into Ise Bay or were stranded ashore after being chased by a pod of orcas (killer whales). They occasionally sighted a pod of orcas driving whales to the shore and were well aware of the strength of orcas. In fact, *shachi*—orcas or imaginary creatures based on orcas—became a symbol of power for feudal lords so that statues of *shachi* called *shachi-hoko* were installed on the rooftops of their castles as *shibi* (ornamental tiles).[7]

Shachi-hoko were also considered guardians of the castles—somewhat similar to gargoyles in Western architecture—because they were believed to blow out water from their blowholes and bring rain. (Most Japanese houses were made of wood and paper, and thus fire was intensely feared.) For example, a pair of gold *shachi-hoko* adorn the rooftop of the majestic central tower of Nagoya Castle in Aichi prefecture, which was originally built in 1609 by Tokugawa Ieyasu for his ninth son, Tokugawa Yoshinao. Nagoya Castle was totally destroyed in May 1945 during the massive air raids of Nagoya by US B-29s during World War II. It was rebuilt in October 1959.[8]

Classical Whalers in Mikawa Province

Mikawa province (the eastern part of Aichi prefecture) suffered the same geographic disadvantages as Owari province. Also, the official history of classical whaling in Taiji mentions that the Owari-Senga Whaling Guild did not adopt the net method invented by Wada Kakuemon Yoriyharu in Taiji in 1675 (see Chap. 6), but adhered to the old harpoon method. As with the case of the Daigo Whaling Guild in Awa-Katsuyama in current southern

Chiba prefecture (see Chap. 4), none of the classical whaling guilds east of Owari province adopted the net method, whereas those west of Kishū, such as on Shikoku and Kyūshū Islands, learned the net method from Waka Kakuemon Yoriyharu and prospered. These facts might explain the absence of whale graves and monuments in Shizuoka prefecture and Aichi prefecture.[9]

Monuments for Dolphins in Shizuoka Prefecture

Surprisingly, there are seven monuments for dolphins along the coast of the Izu peninsula. Starting from east to west on the peninsula, two monuments are located in Kawana, Itō, which used to engage in dolphin fishing (dolphins were considered species of fish). One was created in Mishima Shrine by the Kawana-ura Fisheries Cooperative Union in 1923. The size of the monument is 6.7 feet in height × 2.9 feet in width, excluding the stone foundation, and the inscription reads: "*Iruka-ryō kinen-hi*" (Memorial Stone for Dolphin Fishing).[10]

In the neighborhood stands another memorial stone in Sōtō Zen Buddhism Kaizō Temple. This monument, with a size of about 6.4 feet in height × 2.6 feet in width, excluding the foundation, was created in October 1977. The inscription reads: "*Iruka-gun-rei kuyō-tō*" (Memorial Stupa for the Souls of Dolphins).[11]

Going down the peninsula, the oldest existing memorial for dolphins in this region stands in Inatori, Higashi-Izu-chō. The memorial stone was made in 1828, and the inscription reads: "*Iruka-rei kuyō-tō*" (Memorial Stupa for the Souls of Dolphins). Passing Cape Irō and moving to the west coast of the Izu peninsula, three memorial stones for dolphins stand in Arari-ura in Nishi-Izu-chō: the oldest one was erected in 1882, another in 1935, and the newest in 1949.[12]

Going further north on the peninsula one finds the seventh monument in Toi-chō, Izu. This one was created in Ōyabu Benten Shrine in 1959. The existence of the seven monuments for dolphins on the coast of the Izu peninsula suggests that many pods of dolphins had migrated off the peninsula and that dolphin fishing was part of local fishing communities' traditions.[13]

* * *

Hokuriku Region

Leaving the coast of the Pacific Ocean and moving to the Hokuriku Region on the coast of the Sea of Japan—encompassing current Niigata, Toyama, Ishikawa, and Fukui prefectures—there was no major classical whaling base in this region and fishermen only engaged in "passive whaling"—catching stray whales that were stranded ashore in their bay areas. Still, the local fishermen felt sorry for the whales and made graves and monuments for them.[14]

Whale Graves and Monuments in Niigata Prefecture

Whale Grave in Maze, Niigata

There are several whale graves and monuments in Niigata prefecture. Going down from north to south on the coast of the Sea of Japan, there is a grave for a 44-foot-long fin whale that had washed ashore of Maze, Iwamuro village, Nishi-Kanbara county (current Maze, Nishikan ward,), in 1934. The locale is famous for a hot spring, Iwamuro Onsen, by the Maze-Shimoyama Beach. The villagers made a grave for the whale at Kaiun Temple.[15]

"Sacred Skull of Whale" in Nozumi

In turn, Shingon Buddhism Kaiun-zan Rōsen-in Saishō Temple in Teradomari-Nozumi (current Nozumi, Santō county) enshrines the skull of a whale. Local fishermen had the tradition of donating whale skulls at Saishō Temple in dedication to the Konpira God (the Guardian God of Maritime Traffic). Today, the only one skull is preserved as the "Sacred Skull of Whale" in the Treasure House of the temple. This venerable temple, founded in 733 by one of the Four Saints of Tōdai Temple in Nara, Saint Gyōki (668–February 749), also exhibits a mummy of a sacred animal that appears in Japanese mythology as "*Raijū*" (*lit.*, "thunder beast"). The thunder beast could be a cat, weasel, or wolf. Some observers even think that it could be a *hakubishin* (masked palm civet), an animal indigenous to southern China and Southeast Asia. Since this animal was rare in Japan, the Japanese may have considered it to be a mysterious "thunder beast." However, the mummy at Saishō Temple is most likely to be that of a cat.[16]

More significantly, Saishō Temple houses the oldest existing *sokushin-butsu* (a "mummy Buddha"; a Buddhist monk practices his meditation in a sitting posture until his death, upon which the monk is considered to have become a Buddha, and the body is preserved in the sitting posture forever). This mummy Buddha is said to be of Kōchi-hōin (Saint Kōchi, 1290s–1363), who mastered the austere discipline of Saint Kūkai (Kōbo-daishi, see Chap. 2) and died on October 2, 1363 (on the lunar calendar). Since then, the temple has conducted a memorial service for Saint Kōchi, every October. A recent X-ray test proved that the mummy was truly that of Saint Kōchi.[17]

In 2009, renowned scholar of Japanese literature Donald Keene (b. June 1922) "revived" a biography of Saint Kōchi. Keene moved permanently to Japan in September 2011 following the East Japan Massive Earthquake and Tsunami of March 2011, in order to give a moral boost to the Japanese (his legal naturalization took effect exceptionally quickly in March 2012). He was instrumental in rediscovering the book of the *jōruri* puppet play, *Echigo-no-kuni Kashiwazaki Kōchi-hōin go-denki* (Biography of Saint Kōchi of Kashiwazaki, State of Echigo). It had been lost (it was smuggled out of Nagasaki in 1692), but was found in the British Museum in England in 1962. This led to the revival performance of the *jōruri* puppet play in Kashiwazaki in 2009, after a 325-year hiatus. The revival performances were also conducted in London, England. In turn, a confectionary company based in Kashiwazaki, Bourbon Corporation, founded the Donald Keene Center Kashiwazaki in September 2013, where a replica of the study of his former house in New York City was built.[18]

Whale Graves in Teradomari and Monuments in Kahshiwazaki

Elsewhere in Niigata prefecture, there is a whale grave in Matsuzawa, Teradomari-machi, Santō county, made in 1849. In the neighborhood, there is another whale grave in Ōwada, Teradomari-machi. The villagers of Ōwada found a common minke whale (most likely) washed ashore in 1897. They buried the ribs of the whale and made a grave for it in the local temple, Kan'non-dō. Going south to Nishiyama-machi, Kariwa county (current Nishiyama-chō, Kashiwazaki), it is recorded that fishermen erected a memorial stone for a pod of five short-finned pilot whales that had washed ashore in 1956. This monument at Shinzō-in Temple had the inscription:

"*Geigei aiyori kuyō-tō*" (Memorial Stupa for a Pod of Whales). However, the monument has disappeared now.[19]

Further, in Kashiwazaki, local fishermen erected a memorial stone for an 82-foot-long fin whale that had washed ashore in 1910. This monument stands in the compounds of Miyagawa Shrine.[20]

Whale Graves and Monuments on Sado Island

Then, going west to Sado Island, the largest island in Japan on the coast of the Sea of Japan (and the sixth largest in Japan as a whole, after the Four Main Islands of Japan and Okinawa Island). The entire Sado Island was designated as Sado city, Niigata prefecture, in March 2004. Local history records that thirty stray whales were stranded or washed ashore in the bay area on Sado Island in the two hundred years between 1661 and 1861. An accurate record has been kept because the local communities had to report and pay a tax for each whale caught to the local province, even if they were not professional whalers.[21]

Whale Grave in Katano'o

There are several whale graves on Sado Island. For instance, two fishing villages, Katano'o and Tsukibuse, south of Ryōtsu Bay on the eastern side of the island, made a grave and a couple of monuments for a stray whale. In the early morning in February 1861, a fisherman of Tsukibuse found something big floating off Katano'o. A fisherman of Katano'o recognized it as a whale and fishermen of the two villages seized it. It was an about 90-foot-long fin whale and was already dead. The two villages shared the catch and conducted grand funeral services for the whale separately. Both villages erected jawbones of the whale on a hill of as a grave marker. They also erected memorial stupas at the graves. Nevertheless, the jawbone and the memorial stupa in Tsukibuse were stolen during the Taishō period (1912–1926), and only the ones in Katano-o exist now.[22]

The existing monuments for the whale in Katano'o include a jawbone grave marker, a memorial stupa, and a memorial prayer stone. The 4.0-foot-long memorial stupa has the inscription:

> "*Hōshō nenbutsu Hyakuman-ben*" (Offering One-Million-Times Chanting Prayers), 1867[23]

In turn, the memorial prayer stone called *Nenbutsu-ishi* is about 3.3 feet in height, and has an inscription of the typical Buddhist chanting word *Namu Amida-butsu* (Faith in Amida Buddha). These monuments indicate that the villagers in Katano'o commissioned a local Shingon School Buddhism Jizō-in Temple, to conduct an elaborate seventh-year memorial service for the whale in 1867 (according to the traditional counting system) through performing the One-Million-Times Chanting Prayers and erecting the two monuments.[24]

More remarkably, the villagers commissioned the Jizō-in head priest to accord this whale a *kaimyō* (a posthumous "afterlife" Buddhist name given to a deceased parishioner by the temple to which a parishioner belonged. The temple uses the "afterlife" name for the funeral service and the monthly and annual memorial services that follow). Jizō-in Temple was burned in 1904; however, its *kako-chō* (*lit.*, "book of the past"; a death register) of parishioners survived. The death register of the Jizō-in recorded the afterlife name of the whale as follows:

> December 1860 (February 1861 in the Gregorian calendar), in this village, *Kaiō myōō shin'nyo* (King of the Sea, Graceful and Precious Follower of Amida Buddha), the afterlife name of the whale[25]

Since *shin'nyo* denotes a female follower of Buddhism, this posthumous name indicates that the whale was a female, although the title was "King of the Sea." These monuments and documents in the Katano'o community suggest that the villagers treated the whale as if it were a human being, made enormous financial contributions to Jizō-in Temple, and genuinely mourned the death of the whale.[26]

Whale Grave in Shiidomari

In March 1888, a fisherman by the name of Usami Yazaemon, from the Shiidomari village in the Ryōtsu Bay, found a dead 48-foot-long whale floating in the sea after a pod of whales had left the feeding spot. As fishermen from neighboring villages, including Minato, were also fishing in the area and saw the whale, they all claimed the right to take it. In the end, two communities, Shiidomari and Minato, fought for the fishing rights in a local court, which Shiidomari won. The villagers of Shiidomari celebrated the victory, but simultaneously commissioned a local True Pure Land School Buddhism Shinshū-Ōtani Sect Gansei Temple to create a grave for the whale and to accord the whale a *hōgō* (an afterlife name in Pure Land Buddhism). The gravestone has an inscription of the afterlife name of the whale:

> "*Shaku shinsei nōdo geigyo*" (Whale fish follower of Amida Buddha, sounding its voice for chanting prayer).[27]

"*Shaku*" refers to a follower of Amida Buddha in Pure Land Buddhism, whereas "*shakuni*" refers to a female follower of Amida Buddha. Thus, this afterlife name indicates that this whale was a male.[28]

Other Monuments for Whales on Sado Island

There is also a monument for a whale in *Haniu* in Ryōtsu Bay. This was erected for a large whale that ran aground in 1881/1882 and the inscription reads:

> "*Taigei no gyorei-tō*" (Memorial Stupa for the Soul of Big Whale)[29]

More recently, another monument was made for a baleen whale that was stranded ashore in Bentenzaki in Matsugasaki, Hatano-machi on the southern coast of Sado Island in 1985. The local community buried the lower jawbone of the whale and erected a memorial stupa for the whale.[30]

Whale Monuments in Ishikawa Prefecture

Similar to the cases in Niigata prefecture, there were no major classical coastal whaling bases in Toyama and Ishikawa prefectures. Only one bay community, called Unzu, Fugeshi county (current Hōzu county), on the tip of the Noto peninsula in Ishikawa prefecture engaged in classical coastal whaling on a very small scale. Other local fishing communities occasionally took meat from whales that washed ashore in their bays, and there is no record of whale graves or monuments there. What remains in the region are documents consisting of drawings and pictures about classical whaling. For instance, there is a drawing of a simple net method of whaling called "*daiami hogei*" (a kind of fixed net method of whaling) that was drawn during the Kaei period (April 1848–January 1855, in the Gregorian calendar). Also, a picture of a local lord watching whaling entitled "*Tō-kō kujira-tori kanran no e*"

("Picture of Lord Tō Watching Whale Catching") is preserved in the Noto Township Local History Museum.[31]

Whale Graves in Fukui Prefecture

Whale Grave in Hamasaka, Awara

Moving westward on the coast of the Sea of Japan to Fukui prefecture, one finds a whale grave in Hamasaka, Awara, at the northern border of the prefecture (there is a famous hot spring resort in Awara). According to the archival records of Awara Township, in 1812 Hamasaka was a large fishing village of about 700 residents in 170 households. But the bay community had suffered from poor catches of fish and poor yields of rice in previous years, due to unfavorable weather. This led the villagers to storm a local merchant's residence and demand money (a kind of uprising) in 1811. Then, in the beginning of 1812, a big whale was stranded ashore in Hamasaka. The villagers took meat from the whale, saving them from starvation, but they also felt sorry for killing it and so created a grave for its remains.[32]

They also conducted a funeral service for the whale and erected a gravestone for it. The gravestone is of the size of about 2.6 feet in height × 0.84 feet in width × 0.8 feet in depth. The inscription is illegible, but the date of erection reads: "January 11, 1812" (February 23, 1812 in the Gregorian calendar).[33]

Currently, the gravestone quietly stands next to an old pine tree, tucked into the mountain owned by the local Shōjun Temple. Recently, the local mayor planned to build a *hokora* (a small shrine) in order to protect the gravestone from increment weather so that local people could pay proper tribute to the whale grave. However, the head priest of Shōjun Temple declined the request because he was concerned that people would leave cans and garbage behind and the mountain would become littered if many people visited the whale grave.[34]

Folklore Legend of God of Oshima and Whale

The local people in Awara revered whales as gods that brought blessings. For instance, a little south of Awara, there is a folklore legend about a whale and a god of Oshima, an island off Mikuni-chō, Sakai. Mikuni is a fishing town, famous for Tōjinbō, the site of a stunning formation of tall cliffs and rocks on the "ria coast" (a coastal inlet formed by the partial submergence of a river valley), which is designated as a Natural Monument of Japan, as well as a Site of Scenic Beauty of Japan, by the Japanese government. Near Tōjinbō, there is the venerable Ōminato Shrine on Oshima, which is connected to the land by a long walking bridge, painted vermillion in color. Tōjinbō and Oshima are all part of the Echizen-Kaga Kaigan (Coast) Park designated by Ishikawa and Fukui prefectures. According to the folklore legend, one day when a god of Ōminato Shrine was going back to Oshima, the god asked the waves to carry him to the island, but the waves refused. Then, the god asked a whale to carry him to Oshima and it agreed.[35]

Whale Grave in Aoto no Irie

There is another whale grave in Fukui prefecture. This one is in Aoto no Irie, a long inlet of Wakasa Bay, ranging from Ōi-chō to Takahama-chō, Ōi county. This whale grave was created in 1918 by Tobishima-gumi, which was a local civil engineering and construction company, founded by Tobishima Bunjirō in 1883 (current Tobishima Corporation). In July 1918, employees of the Tobishima–gumi were working on the construction of the Obama Line of the Japanese National Railways (JNR, currently Japan Railways, JR). Then, at night, a whale swam onto the shore of Aoto no Irie, chasing a school of sardines, and was stranded. The workers seized the 45-foot-long whale and created a grave for it. Currently, the gravestone stands on the shore of Aoto no Irie between Wakasa-Hongō Station and Wakasa-Wada Station on the JR Obama Line.[36]

Notes

1. Kumano Taiji-ura hogei-shi hensan-iinkai, ed., *Kumano no Taiji: Kujira ni idomu machi* (Taiji, Kumano: Town That Challenged Whales), Tokyo: Heibonsha, 1965, 12.
2. Ibid.
3. Ibid., 19–20, 68–69.
4. Ibid.
5. Ibid., 68–71, 83, 187.
6. Yoshihara Tomokichi, “Kujira no haka” (Graves of Whales), in Tanigawa Ken’ichi, ed., *Nihon minzoku bunka shiryō shūsei* (Collection of Documents on Japanese Folk Culture), Vol. 18, Tokyo: San’ichi-shobō, 1997, 435.

7. Kumano Taiji-ura hogei-shi hensan-iinkai, 68–69; "Nihon no kujira bunka: Kujira ni kakawaru densetsu · itsuwa (2)" (Japanese Whale Culture: Folklore Legends and Anecdotes Regarding Whales [2]), http://www.geocities.co.jp/NatureLand-Sky/3011/33kujira-densetu2.html, accessed December 23, 2015.
8. Ibid.
9. Kumano Taiji-ura hogei-shi hensan-iinkai, 19–20, 82–83.
10. "Kawana no iruka-hi 2-ki" (Two Memorial Stones for Dolphins in Kawana), July 7, 2010, http://www.asahi-net.or.jp/~rn2h-dimr/ohaka2/12kujira/14cyubu/izuiruka_kawana.html
11. Ibid.
12. Ibid.
13. Ibid.
14. "Kujira bunka: Kujira o tomuratta kujira-baka · kujira-zuka nado" (Whale Culture: Whale Graves and Monuments to Mourn the Deaths of Whales), http://www.geocities.co.jp/NatureLand-Sky/3011/kujirahaka.html, accessed November 14, 2015.
15. Yoshihara, 442–443.
16. "Kujira bunka: Kujira o tomuratta kujira-baka · kujira-zuka nado"; "Saishō-ji" (Saishō Temple), http://www.saisyouji.jp, accessed December 10, 2015.
17. "Saishō-ji."
18. Ibid.; "The Old Puppet Jōruri: The Tale of High Priest Kōchi," June 2, 2017, http://www.bl.uk/events/the-old-puppet-jōruri-the-tale-of-high-priest-kōchi-02-june-2017; "Ko-jōruri Ei e maimodoru" (Old Jōruri Returns to England), *Tokyo Shimbun*, June 3, 2017.
19. "Kujira bunka: Kujira o tomuratta kujira-baka · kujira-zuka nado."
20. Ibid.
21. Yoshihara, 443–445.
22. Ibid., 445; "Kujira bunka: Kujira o tomuratta kujira-baka · kujira-zuka nado."
23. Yoshihara, 445–446.
24. Ibid.
25. Ibid.
26. Ibid.
27. Ibid.; "Kujira bunka: Kujira o tomuratta kujira-baka · kujira-zuka nado."
28. Ibid. (both).
29. "Kujira bunka: Kujira o tomuratta kujira-baka · kujira-zuka nado."
30. Ibid.
31. Yoshihara, 446–447.

32. "Kujira bunka: Kujira o tomuratta kujira-baka · kujira-zuka nado"; "Kujira-zuka" (Whale Grave), "http://www.big-c.or.jp/~makichan/0kuziratuka.html, accessed December 17, 2015.
33. Ibid. (both).
34. Ibid. (both).
35. "Kujira-zuka."
36. "Kujira-zuka"; "Tobishima Kensetsu" (Tobishima Corporation [its official English name]), https://www.tobishima.co.jp/company/company_ayumi.html, accessed December 17, 2015; "Tango no densetsu 45-shū: Kujira-tori" (Folklore Legends in Tango No. 45: Whaling), http://www.geocities.jp/k_saito_site/bunkn45.html, accessed December 25, 2015.

CHAPTER 6

Sacred Mountains of Buddhism and Shintoism and Whaling Culture in the Kansai-Kinki Regions

This chapter examines whale shrines and temples in the Kansai-Kinki Region (in central-western Honshū)–encompassing such prefectures as Mie, Wakayama, Osaka, and Kyoto. This chapter specifically features memorial rites and monuments for whales in Kumano in the Nanki Region (southern part of the Kii peninsula), where Japanese classical whaling is said to have originated, as well as those in Taiji, the seat of one of the four major classical whaling bases in Japan, the Kishū Classical Whaling Base.

Whale Shrines and Temples in Mie Prefecture

Whale Graves in Ise

Beginning with the northeastern coast of the large Kii peninsula, there is a whale grave in Ise on the Shima peninsula. Ise is the seat of one of the most sacred shrines in Japan, Ise Grand Shrine, while the Shima peninsula constitutes a major part of the Ise-Shima National Park, adorned with scenic cliffs, rocks, and islands. On the Shima peninsula, for instance, the famous "*Meoto-iwa*" (Married Couple Rocks) at Futami-Okitama Shrine stand off the shore of Futami-chō, Watarai county (Futami township was incorporated into Ise, in November 2005). In April 1897, local fishermen in Futami caught a whale that drifted into the bay at a time when the village was in dire financial difficulties after having exhausted every resource for the first Sino-Japanese War (July 1894–November 1895). The villagers sold the whale and the

M. Itoh, *The Japanese Culture of Mourning Whales*,
https://doi.org/10.1007/978-981-10-6671-9_6

profit saved the village. In appreciation, using part of the profit, the villagers created a grave for the whale and prayed for its soul.[1]

The gravestone is about 2.2 feet in height and its inscription (the Chinese character used for the whale on the inscription indicates that this whale was a female) reads:

> "*Toku-gei-shi kuyō-tō*" (Memorial Stone for the Whale Caught)[2]

In February 1903, local fishermen in the Shō district of Futami caught a whale that was chasing a school of sardines and was accidentally trapped in a dragnet on the beach. With Futami being part of the "territory of the gods" of Ise Grand Shrine, local villagers had been heavily taxed for the cost of the "Transfer of Ise Grand Shrine." Every twenty years the two main shrines and fourteen other shrines, as well as the treasure house and other buildings, of Ise Grand Shrine are rebuilt on a new lot next to the existing one. This is done in order to transfer each god to a new shrine for the purpose of cleansing the spirit. To resolve these financial difficulties, the villagers sold the whale and allocated the profit for the Transfer of Ise Grand Shrine. In appreciation and in order to pray for the soul of the whale, the villagers created a grave for the whale and erected a gravestone.[3]

It is about 1.7 feet in height × 1.7 feet in width. The inscription (the Chinese character for the whale indicates that this whale was a female) reads:

> "*Taikai kujira no haka*" (Grave for Whale of the Ocean)[4]

Monuments for Whales in Toba

Going down the Shima peninsula, one arrives at Toba, the site of the world-renowned Mikimoto Pearl Island. The fishing community of Ōsatsu in Toba engaged in classical whaling and has a folklore legend about a whale: Riding on the back of a whale, *Jūichi-men Kan'non-bosatsu* (the Eleven-Faced Goddess of Mercy) arrived in Ōsatsu at a cape sharply protruding into the Pacific Ocean, and then brought the blessings of the sea to the village. Hence the cape is called Kujira-zaki (Cape Whale). In commemoration of this folklore legend, the community of Ōsatsu erected several monuments at Cape Whale, such as the "Memorial Monument for the Original Site of the Eleven-Faced Goddess of Mercy," the "Memorial Monument for Cape Whale," and the "Memorial Monument for the Guardian God of Fishermen." In 1989, a whale was accidentally caught in a dragnet on the beach. The Ōsatsu Township Association erected a memorial stone for the whale on Cape Whale. The inscription read:

> "*Kujira kuyō no tō*" (Memorial Stone for Whale)[5]

Since 1989, the Ōsatsu Township Association has organized an annual Ten'no whale festival in dedication and appreciation to the Whale God. For the festival held in July 2014, the townsfolk created a portable shrine palanquin (*mikoshi*) in the shape of a whale, and about fifty men wearing happi coats carried it from downtown to the Ōsatsu fishing port. Then, following the local belief that returning a whale to the sea would bring a good catch of fish, they floated out the "giant whale shrine palanquin" with the help of local *ama* (female divers who catch marine produce, such as Japanese pearl oysters, sea urchins, and seaweed). Elsewhere in Mie prefecture, the fishing community in Suzuka has performed a whaleboat song at the Ten'no whale festival, while Yokkaichi has conducted its own whale festival.[6]

Whale Shrine and "Whale Stone" in Shima

Going further south on the Ise peninsula to Shima, one finds a whale shrine called the Nakiri Shrine in Nakiri, Daiō-chō at the tip of the peninsula. Being part of the Ise–Shima National Park, Nakiri Shrine is located right next to scenic Cape Daiō, where the Daiō Lighthouse overlooks the Pacific Ocean. At the entrance of Nakiri Shrine, a *hokora* (a small shrine as a repository for the god) for the Hiruko God (the Whale God, see Chap. 2) stands next to a giant stone lantern on the left side of the tall *torii* (sacred entrance gate). This small shrine for the Hiruko God is probably made of stone in order to withstand the inclement weather, as the location is in the path of seasonal typhoons.[7]

On the premises of Nakiri Shrine stand the *Kujira-ishi* (Whale Stone). The local fishing communities engaged in whaling from around 1645 to 1715. An archive at Nakiri Shrine records that whaleboats were created in the locale as early as at the beginning of the Shōhō period (January 1645–April 1647). Then, about three hundred years ago, fishermen in Nakiri village seized a large whale that had been stranded ashore, having been chased by orcas. They were surprised to find a big stone from its belly—in the shape of a huge egg, and with the size of about 1.9 feet in height × 1.3 feet in width. The fishermen then feared that the gods of the

Photograph 6.1 Whale Stone, Nakiri Shrine, Daiō, Mie prefecture, circa 1700

sea might punish them (because the stones found in the ocean were considered sacred) and donated the stone to the local Ebisu Shrine, as a memorial. The shrine made an inscription "Whale Stone" on it.[8]

Later, this whale stone was transferred to Nakiri Shrine in order to be worshiped jointly with other whale stones that had been preserved at the shrine. Currently, several smaller whale stones sit in front of the large whale stone. Also, a small memorial stone has been erected to the side (Photograph 6.1).[9]

Whale Grave in Shiroura, Miyama

Going down the large Kii peninsula further to the south, one arrives at a major city in the region, Owase. Near the city, there is a whale grave in a fishing community, Shiroura, Miyama-chō, Kita-Muro county (which was incorporated into a newly created town, Kihoku-chō, Kita-Muro county in October 2005). This grave was created for a whale caught on December

18, Hōreki 8 (January 5, 1759 in the Gregorian calendar) and stands in front of Sōtō-School Zen Buddhism Shōkei-zan Jōrin Temple. The local whaling guild made a sizable financial contribution to Jōrin Temple and commissioned the head priest to bury the whale, with an elaborate funeral service, and also to erect a memorial stupa for the whale at the grave. The stone monument is about 1.6 feet in height and the inscription reads:

> "*Harako-mochi-kujira bodai-no-tō*" (Memorial Stupa for the Whale Carrying a Fetus to Attain the Enlightenment of Buddha) and "In subscription to the greatly merciful and compassionate *Kan'non-bosatsu*" (the Goddess of Mercy)[10]

The magnanimous inscription suggests the depth of the sorrow of the local whaling guild. The inscription ends with the names of two people who made financial contributions to the temple for erecting the memorial stupa—the foreman of a whaleboat, Uematsu Seisuke, and a principal harpooner/*hazashi*, Muradayū, representing the local whaling guild—and the date of the death of the whale, "December 18, Hōreki 8."[11]

Memorial Services for Whale in Shiroura, Miyama

In addition, the guild members commissioned the Jōrin Temple head priest to make an *ihai* (Memorial tablet) for the whale and conduct a funeral service for it. (This memorial tablet has been preserved at Jōrin Temple.) The inscription on the memorial tablet reads:

> "*Harako-mochi-kujira no shōryō-den*" (For the Soul of the Whale Carrying a Fetus)[12]

Before the funeral service for the whale, the whaling guild members cleaned their whole houses inside and out, and also cleansed themselves in the ocean for the purpose of purifying the body and the spirit. Then, they sent out a *segaki* ("feeding the hungry ghosts") boat to the fishing grounds in the sea (*segaki* is a funeral service for those who have no living relatives to take care of their afterlives and to pray for their souls in order to stop their sufferings, see Chap. 3). As if the funeral service were not enough, the guild members held additional memorial services for the whale on its three "monthly memorial days": January 18, May 18, and September 18, 1759, in the lunar calendar. Since the whale died on December 18, Hōreki 8, the days that fell on the 18th of a month were the monthly memorial days of this whale.[13]

The guild members held the one-year anniversary memorial service on December 18, 1759 in the lunar calendar. The light in front of the most sacred statue of Jōrin Temple was lit all day long during these services. Moreover, the guild made a separate financial contribution in the amount of 2 ryō to Jōrin Temple, so that the temple would conduct memorial services for the whale forever. Further, remarkably, two hundred years later in 1957 (by the old counting system), the local community of Shiroura repaired the whale grave that had been ruined over the course of two centuries. They also created a new small shrine, called Kaikō Shrine, next to the grave and erected a new monument for the whale. The inscription read:

> Memorial Stone in Commemoration of Two Hundred Year Memorial Anniversary[14]

The community also commissioned a new memorial tablet for the whale, and thus two almost identical memorial tablets for the whale exist now. In addition, the local communities in Miyama, Kusu-chō, and Yokkaichi have conducted whaleboat memorial services annually.[15]

Legend of Whale Spirit in Shiroura

The inscription for the new memorial stone erected in 1957 also has a detailed explanation as to why the local whaling guild of Shiroura made such enormous financial contributions to Jōrin Temple for the whale carrying a fetus in 1759. It reads:

Classical whaling in Japan originated in Kumano-Muro, and Shiroura was one of the bases.... Whales were the kings of the sea and people revered them as the Ebisu God. On the eve of December 18, 1758, Jōrin Temple seventh head priest, Sōhō-Setsumon, had a strange dream in which the spirit of a pregnant whale appeared and begged him not to catch her until she gave birth. Nevertheless, before the head priest had a chance to tell of it, the whaling guild had already sailed at dawn and caught a large North Pacific right whale, which turned out to be carrying a fetus. Then, a vengeful spirit struck the fishing village of Shiroura, causing an epidemic of plague and no catch of fish, and the community was impoverished. In order to atone for the sin of catching the pregnant whale, the villagers held funeral and memorial services for the whale so that the soul of the whale could attain the enlightenment of Buddha in heaven. Afterwards, the vengeful spirit disappeared and the gentle spring breeze embraced the village....[16]

According to local history, in the dream of Jōrin Temple head priest Sōhō-Setsumon, the whale spirit said, "I am passing through your bay tomorrow on my way to give birth in the South Sea, in order to avoid attacks of orcas. Please ask the whaling guild members not to catch me. If I have a safe childbirth, I will have myself caught in this bay on my way back. On the other hand, if my wish is not granted, your village will be punished." Sōhō-Setsumon realized that the whale caught by the guild was the one that had appeared in his dream, and told the guild head, Uematsu Seisuke, about his dream. Uematsu promised him that he would not catch pregnant whales any longer. However, it was too late.[17]

Uematsu's child and the principal harpooner/hazashi Muradayū's child died on the same day on May 9, 1759 (in the lunar calendar). Then, Muradayū and other adults in the village died of sickness in November. The *kako-chō* ("book of the past"; a death register) of Jōrin Temple in fact recorded that thirty-six people died at that time. The villagers suffered from the loss.[18]

A similar folklore legend of *nidana* (Buddhist concept of "chain of causation") concerning catching a pregnant whale or a mother whale with a calf was told in other classical whaling communities in Japan. The close connection between the Kishū Classical Whaling Base and other classical whaling bases—fishermen in other regions came to Taiji to learn the net method, while natives of Kishū went to other fishing communities to teach the method there—might account for the ubiquity of similar folklore legends in Japan (examined below).[19]

Classical whalers also knew from their own experiences that pods of whales migrated from the north to the south in the fall and then migrated back to the north in the spring after the female whales gave birth in warm waters. The Japanese whalers instinctively felt something special about pregnant whales and mother whales with calves (because this directly concerns reverence for life) and refrained from catching them. Although they did not have modern scientific knowledge, they had an instinctive understanding of the laws of ecology from their own experiences and felt that they should forgo the short-term gain of catching a pregnant whale or a mother whale for the sake of long-term preservation of whale stocks. These lessons took the form of folklore legends and taught them taboos in their conduct.[20]

Memorial Stupa for Whales and Whale Shrine in Owasa

In addition, there is a monument for whales at Fusai Temple in Owase, near the southern tip of the Kii peninsula. It was erected in 1841 and the inscription reads: "*Kujira kuyō-tō*" (Memorial Stupa for Whales). In addition, there is a whale shrine in Kuki-chō, Owasa. In 1764, the local whaling guild members in Kuki Bay donated a pair of stone lanterns to Kuki Shrine. This was made in dedication to the Whale God and with prayers for a good catch, rather than in atonement and with prayers for the souls of whales. According to the inscription on the stone lanterns, local women donated candles for the lanterns in November 1845. In addition, the community of Owase has held a whale festival annually, while the neighboring Shingū (in Wakayama prefecture) has performed a whale dance.[21]

Memorial Stupa for Thirty-Three Whales in Kumano

Going further down the Kii peninsula, there is another notable monument for whales in a small fishing community, Nigishima (not Nikijima), Kumano. Nigishima used to be an island ("*shima*" means island in Japanese) and the area was surrounded by the sea. Historically, it has been said that Japanese classical coastal whaling originated in Kumano and all of the local bay villages in the area engaged in classical whaling. For instance, the local history records that one of the bay villages in Kumano caught a 48-foot-long whale as early as 1035 (Chōgen 8) during the Heian period and obtained about 1430-gallons of whale oil from it. Then in 1686, a local whaling guild built a whaleboat with financial assistance from Kii province and thus began classical whaling operations in earnest. In 1907, the Japan Whaling Company created a whaling base in Nigishima, which operated until 1915.[22]

In addition, Kumano is also known as a sacred place for Shintoism as well as for Buddhism (a manifestation of the syncretism of Shintoism and Buddhism, see Chap. 2). Specifically, the Nigishima Pass is part of the Ise Route of the Kumano Ancient Road, which is the pilgrimage route that connects the three Kumano Grand Shrines and Ise Grand Shrine. The area constitutes the "Sacred Sites and Pilgrimage Routes in the Kii Mountain Range," which was designated and registered as a UNESCO World Heritage Site in July 2004, and attracts many pilgrims and tourists today. These sacred sites include Mount Kōya, in Wakayama prefecture, where Kongōbu Temple, the supreme head temple of Kōya-san Shingon School Buddhism, founded by Saint Kūkai (774–April 835), is located.[23]

A venerable memorial stone for whales stands at the foot of Nigishima Pass leading to Sengen Shrine, not far from the current JR (Japan Railways) Central Nigishima station. This granite monument stands tall more than 4.3 feet high and is in the style of *kaku-tōba* (memorial stupa in the shape of a cube) with the top curved in the shape of a pyramid. This monument was commissioned by Kinomoto Hikobee (he appears to be a foreman of a whaleboat), who commissioned a local temple head priest to erect the memorial stupa in April 1671. The inscription reads:

> "*Kujira sanjū-san-bon kuō-tō*" (Memorial Stupa for Thirty-Three Whales)[24]

This monument was created based on the local tradition that fishermen erected a memorial stone each time the number of fish in a catch reached thirty-three. There used to be several other memorial stupas for whales in the neighborhood, but at least four of them erected near the Ai River have been washed away. This monument, made in 1671, is the oldest existing monument for whales in Japan and was designated as a Cultural Property of Kumano City in July 1969.[25]

Memorial Stupa for Whale in Kumano

In addition, there is another monument in Kinomoto-chō, in the central part of Kumano city. In November 1880, local fishermen in Kinomoto caught a large fin whale that had been chased by orcas and ran aground. Using ¥500 out of the total profit of ¥1500, the community built a public school, Kinomoto Elementary School. The community residents simultaneously mourned the death of the whale and erected a memorial stone for it at the shore near the school, in order to pray for its soul. It is about two feet high and its inscription reads: "*Geigyo kuyō-tō*" (Memorial Stone for Whale).[26]

* * *

Whale Shrines and Temples in Wakayama Prefecture

Crossing the border of Mie prefecture and moving to Wakayama prefecture, one finds a famous whaling community near the tip of the Kii peninsula: Taiji-chō, Higashi-Muro county. Taiji has been known to Western observers of modern Japanese whaling for some time and was the target of attacks by extreme environmental protection groups such as the Sea Shepherd Conservation Society. The reputation of Taiji worsened as a controversial documentary film about the net-drive catching of dolphin

(The Cove, 2009) was widely distributed around the world. The secret filming, done on location without obtaining permission from the township or consent from the fishermen, caused much debate.[27]

Despite its international reputation, Taiji is actually a rather small town. Historically, Taiji was a little fishing village, protruding into the Pacific Ocean, and villagers lived at subsistence level. The terrain of Taiji did not suit the cultivation of rice, the Japanese staple, and the fishermen lived on barley and potatoes. The villagers sustained themselves by fishing and occasionally caught whales that migrated off Taiji Bay.[28]

History of Classical Whaling in Taiji

Classical coastal whaling began in earnest in Taiji in 1606 when Wada Chūbee Yorimoto organized a harpooners' whaling guild, with the assistance of a fisherman from Morozaki in Owari province (current western part of Aichi prefecture) by the name of Denji and an unemployed samurai from Sakai in Senshū (Izumi province, current southwestern part of Osaka prefecture) by the name of Iemon. This is considered to be the first successful organized whaling system in Japan. Then, in 1624, in dedication to the Ebisu God (the Whale God and God of Fisheries), Wada founded a new Asuka Shrine in Taiji through *bunrei-kanjō* ("division and propagation of the 'souls of the gods'"; like propagation of plants by dividing the roots) from Asuka Shrine in Shingū in Kumano. Upon retirement, Wada assumed the name, "Kin'emon," and the eldest son of the Wada clan in following generations succeeded to the name "Wada Kin'emon."[29]

In 1618, Wada's eldest son, Wada Kin'emon II, Yoriteru, hired a fisherman from Onoura near Morozaki, in Owari province, by the name of Yoheiji, who was a skilled harpooner. Yoriteru gave Yoheiji a position of "*hazashi*" (*lit.*, "blade thruster"). This is said to be the origin of the job title name, *hazashi*. Yoriteru's elder son, Wada Kin'emon III, Yorioki, was not very healthy and gave his younger brother Wada Kakuemon Yoriharu the authority to manage the whaling guild, and he himself became an advisor.[30]

In 1675, Wada Kakuemon Yoriharu devised the net method (*amitori-hō*) and combined it with the manual harpoon method. This revolutionized Japanese classical whaling, because Wada did not monopolize his invention, despite the norms of the traditional guilds during the Edo period. Specialized skills and techniques in the guilds were exclusive properties of the guilds, by order of the province, and giving them away to outsiders was illegal. Nevertheless, Wada was willing to give away his method to whalers in other regions because he felt that people should share the blessings of the sea. In turn, his family and senior guild members strongly opposed this, so the issue was brought to the local *karō* ("house elder"; advisor to the province lord) of Kishū province, Mizuno.[31]

Mizuno was open minded and concurred with Wada Kakuemon Yoriharu. Thus, Wada not only taught his method to Tada Kichiemon, the whaling guild head in Muroto-Tsuro in Tosa province on Shikoku Island, but also volunteered to loan his own men—ten *hazashi* (blade thrusters) and sixty *kako* (crewmen)—as well as several whaleboats to Tada and had them accompany Tada on his way back to Tosa in 1683 (see Chap. 8). For this selfless deed, the second lord of Kishū province Tokugawa Mitsusada gave Wada Kakuemon Yoriharu the surname "Taiji" and the privilege to carry samurai swords. Thus, Wada Kakuemon became "Taiji Kakuemon."[32]

Since then, the eldest son of his successive descendants assumed the name "Taiji Kakuemon." Further, Taiji Kakuemon taught his method to whalers in the Saikai Classical Whaling Base on Kyūshū Islands. Thus, Japanese classical whalers in the western part of Japan employed the net method for more than two hundred years until they adopted modern Western methods in the late nineteenth century.[33]

Meanwhile, as the whaling operations of Wada Kin'emon and Taiji Kakuemon grew larger, their guilds developed a fine division of labor among their members, and each member developed specialized skills. For instance, *hazashi* (blade thrusters), who lives were most at risk, regularly trained their bodies during the off-season of whaling, in order to be able to execute their job properly. Also, *funa-daiku* (*lit.*, "boat-carpenter," who specialized in making ships) knew every detail of the crafting/mechanics of whaleboats, including the dynamics of the ship against the water, so that they built sophisticated whaleboats without drawing blueprints first. The painters of whaleboats were also well trained and equipped with specialized skills. They held two paintbrushes in each hand, painted intricate designs and patterns on the boats from scratch without first drawing outlines.[34]

If these professionals did not have an heir of their own, they adopted promising boys and trained them to become their heir (like the tradition of Kabuki actors). These highly specialized skills became the patrimonial property of each family, and they were passed down to successive generations, along with their "house names" (*yagō*). Commoners were not allowed to assume surnames during the Edo period, and they were instead referred to by their "house names," like "Narikoma-ya" and "Nakamura-ya" in Kabuki.[35]

Memorial Rites for Whales at Asuka Shrine and Enkai-in Temple

Since the seventeenth century, local fishermen and whalers in Taiji tried to atone for their deeds of catching whales and prayed so that the souls of whales could attain the enlightenment of Buddha in heaven. Significantly, Asuka Shrine, founded by Wada Chūbee Yorimoto (Wada Kin'emon I) in 1624, has preserved archival documents concerning several memorial rites for the fetuses of whales. They include an account of such a memorial service, in which the *hazashi* (blade thruster) carefully wrapped the fetus of a whale in his own *haori* (a formal kimono jacket) and then buried it. The documents appear to have originally belonged to the Tendai School Buddhism Temple Enkai-in, which used to exist right next to Asuka Shrine. Because of the syncretism of Shintoism and Buddhism, a Shinto shrine and a Buddhist temple often exist next each other.[36]

Enkai-in Temple was one of the local Thirty-six Nachi Gongen Temples ("Gongen" is a local guardian god of Japan, a manifestation of Buddha in the form of an indigenous god as a result of the syncretism of the Tendai School Buddhism and San'nō Shintoism). Nevertheless, due to the suppressions of Buddhism by the Meiji government and the resultant abolitions and consolidations of Buddhist temples into Shinto shrines (see Chap. 2), it appears that Asuka Shrine has "inherited" properties of Enkai-in Temple. In fact, Asuka Shrine has also preserved an Eight-Volume Lotus Sutra that Taiji Kakuemon Yorikatsu donated to Enkai-in Temple in October 1722 (in the lunar calendar).[37]

Gravestone for Whales at Tōmyō Temple

In addition, two local Rinzai School Zen Buddhism temples, Junshin Temple and Tōmyō Temple, are specifically dedicated to praying for the souls of whales and have regularly conducted memorial services for whales to this day. Although both temples belong to the same Myōshin-ji Sect of Rinzai School Zen Buddhism, there is a clear distinction between them in terms of the ranks of their parishioners. Junshin Temple is the "official" family temple for the Wada and Taiji clans, and was originally founded by Taiji Kakuemon II in 1686 for his daughter who died young. A set of 6.6-foot-tall stone statues of the Six *Jizō-bosatsu* (Guardian Deities of Motherhood and Unborn Children; the size is exceptionally large for a statue of

Jizō-bosatsu) solemnly stand in a straight line along the steps leading to the graveyard at the back of the main temple.[38]

In the graveyard, there are the graves of Wada Chūbee Yorimoto (Wada Kin'emon I) and his descendants, as well as of those of people unrelated to the Wada and Taiji clans, who held important positions in their whaling guilds. In contrast, Tōmyō Temple honors common fishermen who worked for the whaling guilds at lower ranks.[39]

In a compound of Tōmyō Temple stands a gravestone for whales. It is 5.0 feet in height × 0.6 feet in width × 0.6 feet in depth. Its inscription reads:

> "*Bōgei-shūrei-tō*" (Memorial Stupa for the Souls of Dead Whales). Wishing that this prayer will bring blessings widely upon us and the whales together so that all of us and the whales can go to heaven and become a Buddha upon death. We have gathered for atonement and offered the prayer of the "precious stone sutra." Erected by Hama Hachibee, Taiji Bay, March 18, Meiwa 5.[40]

The date of erection translates as May 4, 1768 in the Gregorian calendar. The prayer of the "precious stone sutra" refers to a style of Buddhist prayer by copying down words of the sutra on small stones and burying them around the grave of the deceased. Hama Hachibee was a local fisherman and a member of the "Kabachi-gumi" ("*kabachi*" means the head of a whale), which refers to a group of families that had the right to receive a whale's head from each catch. Thus, Hama might have buried a part of a whale's skull at the site (Photograph 6.2).[41]

The Kabachi-gumi appeared to have been organized when the Wada and Taiji clans consolidated local whaling operations and gave local fishermen who had been engaged in whaling the right to whale heads as a form of compensation. The Kabachi-gumi members might have felt a strong need to atone for their deeds because the head of a fish (or animal) was considered the most sacred part of the body. The fact that the local fishermen at the bottom echelon of the whaling guilds, who normally lived at a subsistence level, erected this monument by making a considerable contribution to the temple suggests the depth of their remorse, as well as their sincere desire to atone for their sins.[42]

The gravestone for whales at Tōmyō Temple used to be surrounded by several small stone statues of *Jizō-bosatsu* in different sizes and styles, which were arranged in a rather irregular way. This suggests that fishermen might have buried a small part of a skull under each stone statue at each season.

Photograph 6.2 Whale Grave (1768, left), memorial prayer stone (circa 1800, right), and new stone statue of guardian deity of unborn children for the souls of whales (1986, center), Tōmyō Temple, Taiji, Wakayama prefecture

(The small stone statues of *Jizō-bosatsu* were replaced by a large one in June 1986, which stands next to the gravestone.) Also, Tōmyō Temple erected another monument for whales next to the gravestone during the late Edo period. This is a "memorial prayer stone" made of stone in the shape of an oval, about four feet high. It has the engraving of the six-character *nenbutsu* chanting phrase, "*Namu-Amida-butsu*" (Faith in Amida Buddha), but it is written in a distinctively unique cursive style.[43]

The original calligraphy for the inscription was written by the renowned high priest of Hongan Temple in Kyoto, Saint Tokuhon (July 1758–November 1818), during his visit to Taiji (see Photograph 6.2). Saint Tokuhon is known for his distinctive cursive calligraphy, referred to as the *Tokuhon-moji* (Tokuhon characters), and wrote a similar inscription for memorial prayer stones in temples throughout Japan. Further, at the entrance of Tōmyō Temple stands a graceful stone statue of *Gyoran kan'non* (the Goddess of Mercy for Fish, see Chap. 3).[44]

These records in Taiji suggest that local fishermen who were lower-rank members of the whaling guilds conducted memorial services for whales at Tōmyō Temple, whereas higher-rank guild members (such as *hazashi*) did so at Enkai-in Temple, right next to Asuka Shrine, where the Ebisu God was enshrined.

Ebisu Shrine

Elsewhere in town, there is another whale shrine. Ebisu Shrine in Taiji has a *torii* (sacred entrance gate) made of a pair of whale jawbones in dedication to the Ebisu God. This Whale Gate is mentioned in a famous storybook published in 1688, *Nippon eitai-gura* (an anthology of thirty rags-to-riches stories in a total of six volumes), written by the renowned poet/writer, Ihara Saikaku (1642–September 1693). An episode in Chapter Four of Volume Two of the book describes the life of Wada Kakuemon (Taiji Kakuemon), who perfected the classical net whaling method in 1675. The story refers to the Whale Gate at Ebisu Shrine and reads:

> The Whale Ebisu Shrine has a *torii* made of the bones of the fish. It stands 10 feet tall...[45]

The weathered whale gate stands today, but it is not the original one. The original gate was said to have been made of the jawbones of a North Pacific right whale, but the current gate was made of sperm whale jawbones.[46]

Prayer to Lead Whales to the Pure Land

Remarkably, the whaleboats in Taiji profoundly reflect the local whalers' beliefs and their benevolence for whales. Only a few remains of these boats have been preserved today (examined below), but careful studies conducted by professionals, including architects, art historians, and whale museum curators, have made significant findings. Among the three types of whaleboat, each with different functions, the "*seko-bune*" (chaser boats) especially had strikingly brilliant colorful designs painted on their sideboards. The designs of the paintings were simply too refined just for reducing water friction for speed. The designs were also too sophisticated merely for distinguishing whaleboats of differing ranks.[47]

What is unique about the exquisite designs of the paintings is that they were patterned after objects that appear in pictures of the Pure Land of Paradise (the "afterlife" world in Buddhism). Why did the painters craft various artistic patterns based on the pictures of the Pure Land of Paradise where the souls of the deceased are believed to reside?[48]

For instance, the No. 1 chaser boat is decorated with a design of the "phoenix and paulownia." The Chinese phoenix (*hōō* in Japanese) is a sacred bird in Chinese mythology, whereas the paulownia, also known as the empress tree, is considered a sacred tree in China and Japan. The Japanese use designs of paulownia for their family crests or emblems (the official seal of the Government of Japan is a paulownia crest). The Chinese folklore legend has it that the phoenix will only land on the empress tree, and only when a wise ruler is in power.[49]

The No. 2 chaser boat has the pattern of a "halved chrysanthemum." Being associated with perennial youth in China, chrysanthemums are considered to be a most noble flower in Japan (the chrysanthemum seal is the official seal of the Japanese Imperial Family). In comparison, the No. 3 chaser boat is adorned with patterns of "pine, bamboo, and plum trees," which symbolize longevity. All of the designs and patterns for the chaser boats represent sacred beauty, as depicted in the pictures of the Pure Land.[50]

Taiji-chō Historical Archives curator Sakurai Hayato states,

> These colorful designs of the chaser boats are a manifestation of benevolence and repentance on the part of the Taiji whalers. They painted patterns after the Pure Land of Paradise on their whaleboats so that whales could go to the Pure Land of Paradise upon death. They wanted whales to see the beautiful image

of the Pure Land of Paradise so that they could die in peace, thinking of the Pure Land of Paradise. This was their way to try to atone for their sins of killing whales and to pray for the souls of the whales.[51]

Sakurai points out that the mourning culture of whales in Taiji has roots in the local worship of Mount Nachi (in current Nachi-Katsuura-chō, Wakayama prefecture), one of the most sacred mountains in Japan, which is located less than 10 miles northwest of Taiji. The Great Falls of Nachi and other waterfalls in Nachi were sacred training places for Buddhist monks and practitioners of other religions. Among many sacred places on Mount Nachi, Kumano-Nachi Grand Shrine, which enshrines the Nachi Gongen, is one of the three Kumano Grand Shrines.[52]

In turn, Tendai School Buddhism Seiganto Temple (commonly referred to as Mount Nachi Temple) is the No. 1 Temple (the first station) of the Pilgrimage circuit of the Thirty-three Temples of Saigoku (*lit.*, "western country"; corresponds to current Kansai-Kinki Regions) that worship *Kan'non-bosatsu* (the Goddess of Mercy)—therefore it is also known as the Saigoku Kan'non Pilgrimage. Today, along with these shrines and temples, Mount Nachi and the Great Falls of Nachi constitute the "Sacred Sites and Pilgrimage Routes in the Kii Mountain Range" registered among the UNESCO World Heritage Sites (date of inscription, 2004).[53]

Sakurai Hayato adds, due to the syncretism of Shintoism and Buddhism, people in Taiji lived their daily lives on land and sea being immersed in worshiping gods and goddesses on Mount Nachi. Watching Mount Nachi and the Great Falls of Nachi from the sea, the fishermen in Taiji felt that the sacred mountain, the waterfalls, the sea, and the Pure Land of Paradise, where the Goddess of Mercy was said to live, are all connected. They believed that the sea led to the Pure Land of Paradise. As the fishermen caught whales in the sea, they also prayed that the whales' souls would reach the Pure Land of Paradise.[54]

This view is endorsed by the fact that, after catching a whale, all the chaser boats circled around the whale, and all the crewmen held both hands together, solemnly repeating the chanting of "*Namu-Amida-butsu*" (Faith in Amida Buddha), and praying that the soul of the whale would attain the enlightenment of Amida-Buddha. A similar tradition was observed in the Saikai Classical Whaling Base in Kyūshū (see Chap. 9). It was a daunting task for small Japanese men to catch a huge creature in the sea with only rudimentary tools, and therefore they were elated when they caught a whale. However, the excitement was immediately replaced by remorse and sorrow for killing it.[55]

Taiji Whale Museum

Today, the Taiji Whale Museum exhibits a replica of a life-size No. 1 chaser boat. All of the original whaleboats had disappeared by the beginning of the twentieth century. Only three pieces of the original boats remain: a bow of a No. 2 chaser boat and two sideboards of a No. 5 chaser boat. The *igeta* sign (modelled after the upper structure of a "well"; it looks similar to a hash mark or a sharp sign in music notation), which was a *kae-mon* (an informal emblem of a family used for ornamental purposes, not an official emblem) of the Taiji clan, and a vivid design of bamboo were painted on both sides of the bow. In turn, a colorful pattern of *hi-ōgi* (a fan made of Japanese cypress) was painted on the two sideboards (the reverse side was painted entirely in red).[56]

Fortunately, some of the paintings of classical whaling in Taiji, including *Kujira-bune emaki* (Picture Scroll of Whaleboats), have been preserved. They depict distinctive features of each of the whaleboats, as well as their entire formation in the sea. Through painstaking examination and analysis of the original artifacts, a team of specialists on traditional Japanese boats reproduced all of the different types of whaleboats in Taiji, including chaser boats, towboats, and net boats (to carry nets).[57]

These replicas clearly indicate that the whaleboats were truly works of art, with sophisticated designs and exquisite patterns, which in turn suggests the depth of the fishermen's feelings toward whales. According to Taiji Township Historical Archives curator Sakurai Hayato, fishermen in Taiji caught mainly humpback whales, which are fast swimmers and sink after death, and therefore are harder to catch, in contrast to North Pacific right whales, which are slow swimmers and do not sink after death. The classical net method invented in Taiji enabled them to catch humpback whales, but the endeavor still took enormous courage. As catching humpback whales presented more difficulty, fishermen's respect for the whales grew and their remorse deepened.[58]

These tangible and intangible whaling artifacts concerning the mourning of whales in Taiji suggest that their practice is not a pretense on the part of the whalers in order to justify killing whales. Their sorrow and remorse were genuine. The fishermen were confronted with a dilemma between necessity to kill whales for their living and their belief in Buddhism which taught them not to kill any living things, compounded by their worship of whales as a Shinto god and their awe toward these magnificent creatures (the largest "fish" they knew). Accepting their fate as whalers, the fishermen in Taiji gratefully received catches of whales as blessings of the gods, and sincerely prayed for the souls of whales through various memorial rites.[59]

Folklore Legend in Taiji

There is a cautionary tale about classical whaling in Taiji: "Do not even dream of catching a mother North Pacific right whale accompanied by a calf." The fishermen knew from their experiences that the motherly instincts of female whales were usually very strong, so they knew if they caught her calf first they could catch a mother whale because the mother would never leave its calf. On the other hand, the fishermen also knew that this was a very dangerous proposition because the North Pacific right whale was very brave and the mother would become extremely violent if they tried to catch her calf. For this reason, they did not catch mother–child pairs of North Pacific right whales.[60]

Nevertheless, there is a case in which a whaling guild head broke this rule and caused a tragic accident. As the number of whales migrating off the coast of Japan had drastically decreased in the Meiji period (1868–1912), due to the increase in American and European whaling operations in the Japan Ground, Japanese classical coastal whaling operations became very difficult. The classical whalers in Taiji did not know the reason why the catch had decreased and kept doing the only thing they knew to do at the time of a poor catch: offering extra prayers to the Goddess of Mercy at Cape Tōmyō, where the sentry's whale-watching post stood. However, their prayers were not answered. Consequently, the whaling guilds suffered financially and had enormous debts due by the end of the year (the Japanese tried to liquidate their debts by the end of the year so that they could celebrate the New Year's Day debt free).[61]

Large North Pacific Right Whale Incident

Then came December 24, 1878, a day Taiji people will remember forever. Local fishermen in Taiji found a mother–child North Pacific right whale pair swimming into the bay. The fishermen initially caught the whales in a net, but the whales broke the net and swam away. The fishermen chased them into the ocean in the rain, but a whaling guild head, Wada Kin'emon (a descendant of his namesake, the founder of Japanese classical whaling) decided to give up the hunt as the weather worsened. Nevertheless, his cousin, Taiji Kakuemon (a descendant of his namesake, the inventor of the net method), who was also a guild head, decided to chase after the whales and sent his own whaleboats. They were shipwrecked in the storm and as many as 111 crew members were drowned. The community created a memorial tower for the missing persons in 1879. It was a serious blow to the whaling community and directly contributed to the decline of classical whaling in Taiji.[62]

New Monument and Annual Memorial Service for Whales

One hundred years after the tragic incident, in 1979 Taiji township erected a new monument for whales in Kandori-zaki Park on Cape Kandori overlooking the Pacific Ocean, where sentries signaled to the harbor when a whale was sighted. On top of a large cube-shaped memorial stone lies a replica of a North Pacific right whale as if it were alive swimming in the ocean. The inscription on the memorial stone reads:

> "*Kujira kuyō-hi*" (Memorial Cenotaph for Whales)[63]

The auxiliary monument explains that this cenotaph was created in order to express appreciation for whales that had sustained the lives of people in this fishermen's community and to pray for the souls of whales to so that they would attain the enlightenment of Buddha. Since then, the Taiji hogei O.B. ("old boys")-kai (the retired whalers' society) has sponsored an annual memorial service for whales in front of this monument on April 29 every year, in which the head priests of both Junshin Temple and Tōmyō Temple perform the service together. They hold the service at the temples on rainy days.[64]

In the memorial service for 2014 held at Junshin Temple, Taiji town mayor Sangen Kazutaka stated, "Having been born in a place like an island isolated from the land, our ancestors challenged classical whaling in order to survive, and managed to survive by receiving the blessings of whales. We would like to continue to embrace our appreciation for whales and retain the history of whaling in this community for the sake of the future." The Taiji community also holds a memorial service for fish and whales during the Bon Festival (the week of August 15) every year. On August 16, 2012, this author observed one such memorial service officiated by the head priests of Tōmyō Temple and Junshin Temple (see Photograph 6.3).[65]

Photograph 6.3 Memorial rite of sending off the souls of whales to the sea, Taiji, Wakayama prefecture, August 2012

This memorial rite is called *shōrō nagashi* (sending off spirits to the sea), in which people send off a boat carrying the souls of the dead to the sea so that the souls can return to the Pure Land of Paradise beyond the sea (see Photograph 6.4). This is a distinctive tradition mainly practiced these days in Saga and Nagasaki prefectures in Kyūshū. This memorial rite is different from *tōrō nagashi*—floating paper lanterns on a river in order to send off the souls of the dead— that is widely practiced in many other regions of Japan. Also, a local whale dance, which is performed on August 15 annually, is designated as an intangible folk cultural property of Wakayama prefecture.[66]

Memorial Cenotaph for Whales in Kushimoto

Leaving Taiji and moving west to Kushimoto, a major fishing port city at the tip of the Kii peninsula, one finds Kii-Ōshima. This island used to a modern whaling base of the Nippon Suisan Company until 1962, when it ended

Photograph 6.4 Memorial boat for sending off the souls of whales to the sea, Taiji, Wakayama prefecture, August 2012

operations. In 1908, the Kii Suisan Company donated a pair of lower jawbones of a whale to the Kushimoto Jitsugyō School, and the school erected an entrance gate out of the jawbones. A picture postcard made in 1918 records that the whale gate was 9.0 feet in height and 3.2 feet in width. Then, in January 1940, the Nippon Suisan Company erected a memorial stone in commemoration of the Imperial Year 2600 (which was 1940 in the Gregorian calendar). A tall cenotaph, 9.3 feet in height × 1.3 feet in width, stands on top of thirteen steps and three pedestals. Its inscription reads:

"*Chō-gei-tō*" (Memorial Cenotaph for Whales)[67]

* * *

Monument for Whales in Osaka Prefecture

Whale Bridge in Osaka

Moving up the Kii peninsula on the western coast to Osaka prefecture, there is a unique "whale bridge" at Rinzai School Zuikō Temple in Higashi-Yodogawa, Osaka. Originally founded by Prince Shōtoku (February 574–April 622), Zuikō Temple was renowned, but it burned down in a fire around 1334/1336. Then, a Rinzai School Zen Buddhist monk called Ten'nen restored the temple in 1643, and it was renamed as Zuikō Temple in 1729. A temple document records that in January 1757, Zuikō Temple's fourth head priest, Tanjū-Chinin, visited Taiji, where the local community had been impoverished due to the poor catch of whales. Fishing villages that engaged in classical whaling usually lived at a subsistence level for most of the year, and prospered only during the good whaling season. Taiji was no exception.[68]

Taiji whale guild heads pleaded with Tanjū-Chinin to pray for a good catch of whales. He initially declined the plea because killing living things was against the Five Commandments of Buddhism, but in the end he accepted it as he felt that the good deed of saving the fishermen from impoverishment outweighed the bad deed of killing whales. Tanjū thus prayed for a good catch, and his prayer was immediately answered. In appreciation, the fishermen made a sizable financial contribution of 30 ryō (gold coins) and 18 whale bones to Zuikō Temple.[69]

Still, Tanjū-Chinin felt sorry for the whales and wanted to use the donations to make something by which he could pray for the souls of the whales so that they would rest in peace in heaven. Tanjū-Chinin accordingly made a bridge out of the whale bones and installed the bridge across Kōsai Pond in a compound of the temple. This small bridge of about 18 feet long × 9.0 feet wide is called *Setsugei-kyō* (*lit.*, "snow whale bridge") because of its color. As the whale-bone bridge was fragile and decayed over the course of time, it was repaired four times before World War II. Only the handrail of the repaired bridge was made of whale bones and the rest was made of stone for durability.[70]

Then, in June 1945, the temple and the bridge were burned down in the US Massive Air Raid on Osaka. The bridge was again restored in 1974, with the cooperation of the Taiji community, and the temple was restored in 1984. Then the bridge was repaired again in November 2006. In turn, the neighboring community of Dōjima has performed a whale dance.[71]

* * *

Monument for Whales in Kyoto Prefecture

Whale Shrine and Whale Graves in Aoshima, Ine

Ine-chō, Yosa county, the fishing community on the northwestern tip of the Tango peninsula in Kyoto prefecture, has been known for catching *buri* (yellowtail), but there is a record that the local fishermen also occasionally caught stray whales that swam into Wakasa Bay. They used the net-drive method (that was employed in Kayoi, in current Yamaguchi prefecture) by driving whales into Ine Bay and closing the bay with nets. The bay community of Ine had also retained scores of wooden houses built directly by the sea, which have a unique architectural style called *fune-ya* (lit., boathouse). These houses have built-in boat slipways, storage, and working spaces on the first floor, and residential areas on the second floor. These fishermen's houses became the first of the "Groups of Traditional Buildings of Japan" in 1975 designated by the government for the purpose of preservation of historic architectures.[72]

About 500 feet off from Ine in Ine Bay, there is an island called Aoshima, where a whale shrine, Ebisu Shrine, stands in the woods. This whale shrine houses three graves for whales. The Ine local history records:

In February 1808, local fishermen caught a whale and her calf that had strayed into Ine Bay. The mother whale was already dying (probably having been attacked by orcas). The fishermen tried to separate the calf from the mother whale, but the calf refused to leave its mother's side. Thus, they ended up catching both the mother and her calf. Yet, having observed the strong bond between the calf and the mother whale, the fishermen felt sorry for the calf and could not bring themselves to eat it. Instead they buried the body of the calf on the premises of Ebisu Shrine, erected a gravestone for it, and conducted a funeral service for it.... Therefore, the calf whose grave was made should also be a humpback whale.[73]

This account indicates that the mother whale was a humpback whale, but it is unclear whether the fishermen also refrained from eating the mother whale and so buried its body here as well. It is ambiguous because the Japanese language often does not clarify the subject of a sentence (it is assumed from the context) and also whether a subject or an object is singular or plural (again it is assumed from the context). For instance, a previous study has suggested that the villagers did not eat the mother whale and buried her also; however, judging from the way it is written, it appears that this account only refers to the calf and that the villagers consumed the mother. (A mother whale is too large to be buried.)[74]

In fact, another Ine local history records that the local fishermen caught whales and consumed them, and that each time they caught a whale, they made a financial contribution to the venerable Shingon School Buddhism Nariai Temple in Miyazu, at the foot of the Tango peninsula, about 10 miles away from Ine, and conducted a funeral service there. Nariai Temple is the No. 28 Temple of the Pilgrimage circuit of the Thirty-three Temples of Saigoku. The villagers also held a memorial service for whales at Nariai Temple on January 19 (in the lunar calendar) every year.[75]

Three gravestones for whales stand on the left side on the premises of Ebisu Shrine, above stone steps from the *torii* entrance gate. Why are there three gravestones for whales? The town records do not clearly account for this. No previous studies have explained it either. A careful reading of the inscriptions on the gravestones answers the question. The gravestone on the right is clearly for the calf, as the inscription reads:

> "*Ko-kujira-tō*" (*lit.*, "memorial stone for a whale child")[76]

In turn, the inscription on the gravestone in the middle reads:

> "*Zaitai-geishi-tō*" (*lit.*, "memorial stone for a whale child in the womb"), "January 23, Bunka 5" (February 7, 1808 in the Gregorian calendar)[77]

This inscription suggests that the mother whale also carried a fetus. Further, the inscription on the gravestone on the left reads:

> "*Geitai kyōrei tsuisui*" (Wishing to drive away the evil spirit of the whale carrying a fetus)[78]

This inscription uses the Chinese character for the word whale that specifically refers to a female whale. The fishermen, who had already felt sorry for the calf, may have felt more deeply remorseful when they had found out that the mother whale also carried a fetus. They then may have strongly feared that evil spirits would beset the village, as many legends in other fishing communities in Japan warned of the consequences of catching a mother whale with a calf or a pregnant whale. Therefore, in order to atone for catching the pregnant whale and its calf, the villagers buried the calf, and the fetus, and also a part of the spine of the mother whale, and made three graves. Actually, a part of the spine is placed on the ground, next to the gravestone for the mother whale. Thus, the existence of the three gravestones would make sense.[79]

Notes

1. "Kujira bunka: Kujira o tomuratta kujira-baka · kujira-zuka nado" (Whale Culture: Whale Graves and Monuments to Mourn the Deaths of Whales), http://www.geocities.co.jp/NatureLand-Sky/3011/kujirahaka.html, accessed November 14, 2015.
2. Ibid.
3. Ibid.
4. Ibid.
5. Ibid.
6. "Tairyō negau ō-kujira mikoshi" (Giant-Whale Portable Shrine to Pray for a Good Catch), *Chūnichi Shimbun*, July 15, 2014; Boreal Institute for Northern Studies, ed., *Small-Type Coastal Whaling in Japan: Report of an International Workshop*, Edmonton, Alberta: Boreal Institute for Northern Studies, the University of Alberta, 1988, 56.
7. "Nakiri-jinja" (Nakiri Shrine), http://www.geocities.jp/miniuzi0502/jinjadistant/mie/nakiri.html, accessed November 11, 2015.
8. Ibid.; Kumano Taiji-ura hogei-shi hensan-iinkai, ed., *Kumano no Taiji: Kujira ni idomu machi* (Taiji, Kumano: Town That Challenged Whales), Tokyo: Heibonsha, 1965, 20.
9. Ibid. (both).
10. Yoshihara Tomokichi, "Kujira no haka" (Graves of Whales), in Tanigawa Ken'ichi, ed., *Nihon minzoku bunka shiryō shūsei* (Collection of Documents on Japanese Folk Culture), Vol. 18, Tokyo: San'ichi-shobō, 1997, 428.
11. Ibid.
12. Ibid., 428–429; "Kujira bunka: Kujira o tomuratta kujira-baka · kujira-zuka nado."
13. Ibid. (both).
14. Ibid. (both).
15. Ibid. (both); Boreal Institute for Northern Studies, 56.
16. Yoshihara, 428–429 and 472–473.
17. Ibid., 472–473.
18. Ibid.
19. Ibid.; Boreal Institute for Northern Studies, 62.
20. Ibid. (both).
21. Yoshihara, 429; "Kumano-gaku" (Study of Kumano), October 21, 1012, http://www.city.shingu.lg.jp/div/bunka-1/htm/kumanogaku/article/culturalasset/city/index.html; Boreal Institute for Northern Studies, 56.

22. Yoshihara, 430.
23. "Rekishi kaidō" ("Roads to History"), April 23, 2004, http://www.asahi.co.jp/rekishi/04-04-23/01.htm; "Sacred Sites and Pilgrimage Routes in the Kii Mountain Range," 2004, http://whc.unesco.org/en/list/1142
24. Yoshihara, 429–430.
25. Ibid.
26. Ibid., 430.
27. Kumano Taiji-ura hogei-shi hensan-iinkai, 33–43, 143–144.
28. Ibid.
29. Ibid., 66–70, 187.
30. Ibid., 71–84, 188–189.
31. Ibid.
32. Ibid., 80–82, 188–189; Yoshihara, 435.
33. Ibid. (both).
34. Taiji-chōritsu kujira no hakubutsukan, ed., *Kujira-bune: Katachi to ishō* (Whaleboats: Styles and Designs), Taiji: Taiji Whale Museum, 2011, 35–37, 41–42.
35. Ibid.
36. Ibid., 37–38, 40; Sakurai Hayato, author interview, August 16, 2012.
37. Ibid. (both).
38. Kumano Taiji-ura hogei-shi hensan-iinkai, 142–143.
39. Ibid.
40. Ibid., 142; Yoshihara, 430–431.
41. Ibid. (both).
42. Kumano Taiji-ura hogei-shi hensan-iinkai, 142–144.
43. Yoshihara, 431; "*Kuyō-hi kara tadoru hogei-bunka*" (Whaling Culture Inscribed on Memorial Stones), December 18, 2014, http://www.asahi.com/area/wakayama/articles/MTW20141218310520001.html
44. Ibid. (both).
45. Yoshihara, 432; Kumano Taiji-ura hogei-shi hensan-iinkai, 78–80, "Taiji," http://www.cypress.ne.jp/taiji/6.html, accessed September 3, 2015.
46. Ibid. (both).
47. Taiji-chōritsu kujira no hakubutsukan, 33–36, 41–42; "Kujira-bune: Wazuka na shiryō kara hakkutsu" (Whaleboats: "Excavated" from a Few Existing Documents), *Asahi Shimbun*, January 8, 2015.
48. Ibid. (both).
49. Taiji-chōritsu kujira no hakubutsukan, 36–41; "Kujira-bune: Wazuka na shiryō kara hakkutsu."
50. Ibid. (both).
51. Taiji-chōritsu kujira no hakubutsukan, 36–41; Sakurai, author interview.
52. Ibid. (both).

53. Ibid. (both); "Sacred Sites and Pilgrimage Routes in the Kii Mountain Range."
54. Sakurai, author interview; Taiji-chōritsu kujira no hakubutsukan, 37–41; "Kujira-bune: Wazuka na shiryō kara hakkutsu."
55. "Taiji-chō no rekishi · bunka o saguru" (To Search for the History and Culture of Taiji Township), http://www.town.taiji.wakayama.jp/kankou/sub_01.html, accessed September 18, 2015; "Nihon no kujira bunka: Kujira to shinkō" (Japanese Whale Culture: Whales and Beliefs), http://www.geocities.co.jp/NatureLand-Sky/3011/kujira-to-sinkou.html, accessed November 23, 2015.
56. Taiji-chōritsu kujira no hakubutsukan, 33–34, 42; "Kujira-bune: Wazuka na shiryō kara hakkutsu."
57. Ibid. (both).
58. Ibid. (both).
59. "*Kuyō-hi kara tadoru hogei-bunka*" (Whaling Culture Inscribed on Memorial Stones), December 18, 2014, http://www.asahi.com/area/wakayama/articles/MTW20141218310520001.html
60. Yoshihara, 431, Kumano Taiji-ura hogei-shi hensan-iinkai, 146.
61. Kumano Taiji-ura hogei-shi hensan-iinkai, 144–146.
62. Ibid., 146–155.
63. "Kujira ni kansha o komete kuyō-sai" (Memorial Service for Whales with Appreciation), *Asahi Shimbun*, April 30, 2014.
64. Ibid., Boreal Institute for Northern Studies, 58–59.
65. Ibid. (both).
66. "Nihon no kujira bunka: Kujira to shinkō" (Japanese Whale Culture: Whales and Beliefs), http://www.geocities.co.jp/NatureLand-Sky/3011/kujira-to-sinkou.html, accessed November 23, 2015.
67. Yoshihara, 432; "Kujira bunka: Kujira o tomuratta kujira-baka · kujira-zuka nado."
68. Yoshihara, 432–433; Kumano Taiji-ura hogei-shi hensan-iinkai, 182–183.
69. Ibid. (both).
70. Ibid. (both); "Zuikō-ji · Setsugei-kyō" (Zuikō Temple and Setsugei Bridge), March 16, 2009, http://www.city.osaka.lg.jp/higashiyodogawa/page/0000025984.html
71. "Zuikō-ji · Setsugei-kyō."
72. "Ine-chō" (Ine Town), http://www.town.ine.kyoto.jp, accessed December 10, 2015.
73. Yoshihara, 448; "Kujira bunka: Kujira o tomuratta kujira-baka · kujira-zuka nado."
74. Ibid. (both); "Tango no densetsu 45-shū: Kujira-tori" (Folklore Legends in Tango No. 45: Whaling), http://www.geocities.jp/k_saito_site/bunkn45.html, accessed December 10, 2015.

75. Ibid. (all three).
76. Yoshihara, 448; "Kujira bunka: Kujira o tomuratta kujira-baka · kujira-zuka nado."
77. Ibid. (both).
78. Ibid. (both).
79. Ibid. (both).

CHAPTER 7

Pure Land Buddhism and Whaling Culture in the Chūgoku Region

The Chūgoku Region, the western part of the Japanese main island Honshū, is divided into two subregions, each with distinctive climates and cultures; the Sanin Region on the coast of the Sea of Japan and the Sanyō Region on the coast of the Seto Inland Sea. In the Sanin Region, there were no classical whaling bases in Tottori and Shimane prefectures, and there are few records of memorial rites for whales there, with one exception, whereas Yamaguchi prefecture offers a totally different story.

Whale Grave in Shimane Prefecture

While documentation is scarce for monuments for whales in Tottori and Shimane prefectures, there is a whale grave in Izumo, in Shimane prefecture. Izumo is the seat of one of the most sacred Shinto shrines in Japan, Izumo Grand Shrine (formerly called Kizuki Grand Shrine), where Ōkuninushi, one of the principal gods in Japanese mythology and the god of nation building and farming, is enshrined. A whale grave exists in a local Pure Land School Buddhism An'yō-ji Temple, in Kizuki-Kita, Taisha-chō. A whale grave, as well as a stone statue of "*Sakana-jizō*" (the Guardian Deity of Unborn Children for Fish), stand in the compounds of An'yō Temple. Although the provenance of the whale grave is unknown, these monuments suggest that this monzen-machi (town around temple or shrine) of Izumo Grand Shrine used to be a fishing community. An'yō Temple also houses a statue of *Kan'non-bosatsu* (the Goddess of Mercy) that the founder of Kabuki-style dance, Izumo no Okuni (b. ca. 1572, date

M. Itoh, *The Japanese Culture of Mourning Whales*,
https://doi.org/10.1007/978-981-10-6671-9_7

of death unknown; originally a *miko*, a female servant-dancer, at Izumo Grand Shrine), worshiped upon retirement at the Renga Temple nunnery.[1]

In addition, there is a large whale *ema* (*lit.*, "picture horse"; a wooden plaque, originally with a picture of a horse, used in shrines for the purpose of offering prayers and making wishes, see Chap. 2) at Itsukushima Shrine in Setogashima, Hamada, on the western edge of Shimane prefecture. The shrine is generally referred to as the Seto-myōjin. The local whaling guild commissioned a painter by the name of Okada Ittoku of Banshū (current Hyōgo prefecture) to draw a picture of whaling on a large wooden plaque and donated it to Itsukushima Shrine in October 1854. This respectable whale *ema* is of the size of 4.5 feet in length × 6.0 feet in width and depicts a scene of dozens of whaleboats surrounding three whales in Setogashima Bay. This whale *ema* is a valuable picture that indicates local fishermen had engaged in classical whaling, and was designated as a Cultural Property of Hamada City in November 1969.[2]

Whaling Culture and Monuments in Yamaguchi Prefecture

Chōshū province (the current Yamaguchi prefecture) was the seat of one of four Japanese classical whaling bases, the Chōshū Kitaura Classical Whaling Base ("kitaura" literally means "northern bay"). An influential samurai lord of the Mōri clan ruled the domain of Chōshū province—which comprised the State of Nagato and the State of Suō—with the capital in Hagi on the coast of the Sea of Japan. (With historically valuable architecture and monuments, Hagi is a popular tourist destination today.) The Mōri clan encouraged whaling, which brought sizable revenue for the province, and appropriated revenue from whaling for defense of the province when Western countries threatened to break the Japanese isolationist *sakoku* policy and open its ports.[3]

Chōshū province exercised a certain independent power from the Tokugawa shogunate government and became one of the four dominant provinces that were instrumental in overthrowing the Tokugawa government in favor of restoring the power of the emperor and thereby establishing the Meiji government in 1868: the Meiji Restoration. The founding fathers of the Meiji government from Chōshū province included Itō Hirobumi (October 1841–October 1909), who became the first prime minister of

Japan, and Kido Takayoshi (August 1833–May 1877), who held the dual positions of home minister and education minister.[4]

Mishima

Going from east to west on the coast of the Sea of Japan, there are about two dozen whale graves on Mishima, an island about 28 miles northwest from Hagi. In the late seventeenth century, Fukazawa Yogorō (also known as Matsushima Yogorō), a member of the famous whaling guild in Yobuko (in current Saga prefecture) and Ikitsuki (in current Nagasaki prefecture) on Kyūshū Island (see Chap. 9), began classical whaling on Mishima. Local history records that Fukazawa created a total of a thousand graves on the hill of Mt. Takami on the island for the whales that he caught. Another record states that Fukazawa made ninety-five graves for whales on the hill of Mt. Takami at some point in time. These whale graves were made during the Shōtoku period (June 1711–August 1716). While most of the whale graves have been weathered away over the course of time, some of them have withstood the inclement weather and about two dozen still stand at the foot of the mountain.[5]

Nagato

Travelling west from Hagi, one arrives at Nagato, Yamaguchi prefecture. Nagato stands out in terms of the Japanese culture of mourning the death of whales. The local people have retained this tradition in a most religious fashion, so that the renowned anthropologist Nakamura Ikuo refers to Nagato as "the mecca of the culture of mourning the death of whales." What distinguishes Nagato from other fishing communities in Japan that had engaged in classical whaling?

History of Classical Whaling in Nagato

The Sea of Japan was one of the migration routes of whales, and local fishermen in Nagato caught whales that had drifted into their bays. From the mid-seventeenth century to the early twentieth century, five fishing communities in Nagato engaged in classical whaling: Kayoi (Kayoi-ura), Senzaki, Kiwado, Tsuō-Tateishi, and Kawajiri, in order from east to west. There was the sixth one, Misumi (located between Hagi and Nagato), but it was short lived and is little documented. The five communities combined constituted the Chōshū Kitaura Classical Whaling Base. This was the smallest of the four major Japanese classical whaling bases, with the other three being the Kishū Classical Whaling Base in Taiji, the Tosa Classical Whaling Base on Shikoku Island, and the Saikai Classical Whaling Base on Kyūshū Island. Local whalers in Nagato developed a more passive drive-net method (driving whales into a bay and closing the bay with nets), as compared to a more active net method (catching whales with nets in the sea) that was invented in Taiji.[6]

Then, as American and European whalers actively engaged in their operations in the Japan Grounds, whales stopped migrating off the coastern region of the Sea of Japan, and the local whaling guilds in Nagato ended their operations by 1908. Thus classical whaling in Chōshū Kitaura became "extinct." The whaling operations lasted for 235 years in Kayoi (from 1673 to 1908) and in Senzaki (from 1672 to 1907), in Kiwado for 213 years (from 1690 to 1903), in Tsuō-Tateishi for 204 years (from 1699 to 1903), and in Kawajiri for 210 years (from 1698 to 1908). Nagato meanwhile retained its pristine waters and scenic landscape such that the area is now designated as the Kita-Nagato Coast Associate National Park (***kokutei-kōen***; not a full national park, ***kokuritsu-kōen***) by the Japanese government. The official "city color" of Nagato is crystal blue.[7]

Kayoi

Kayoi-ura (Kayoi bay village, "Kayoi" hereafter; incorporated into Nagato city in March 1954) is a 10-mile-long coast on the southeastern tip of Ōmi (*lit.*, "blue sea") Island, with a circumference of about 25 miles, in the Sea of Japan. The northern coast of this island, adorned with strikingly shaped cliffs and rocks, is referred to as the "Ocean Alps." It is designated as a Site of Scenic Beauty of Japan (***kokutei-meishō***), as well as a Natural Monument of Japan (***kokutei-ten'nen kinenbustu***) by the Japanese government. The writer Tsunoda Fusako described her impression of the island: "The harmony of the color of the sky and the sea was magical. The symphony of the resounding waves splashing on the rocks was almost religious." The scenic beauty also impressed the renowned artist of Japanese-style painting Higashiyama Kaii, inspiring him to depict the rocks of Ōmi Island on the walls of the new Imperial Palace in Tokyo. A ferry from Senzaki used to be the only access from Ōmi Island to the main island, Honshū, but the Ōmijima Bridge currently provides easier access to Senzaki.[8]

The origin of classical whaling in Kayoi dates back to 1673 when a member of the local gentry, Hayakawa Seiemon (Seibee), organized the Kayoi Whaling Guild, consisting of about 300 members, and Chōshū province authorized his operation. The operation lasted until 1908 when the guild was disbanded. At that time, the local fishermen sold all of their whaleboats, but kept their whaling tools as precious relics that had sustained the livelihood of the community for more than two centuries. They kept them even during World War II when the Japanese government forced citizens and corporations to "donate" metals, by hiding them in the walls of their houses and covering them with plaster. Consequently, many tools have been preserved to this day, some of which are 320 years old. Among them, 140 items were collectively designated as Important Tangible Folk Cultural Properties of Japan by the Japanese government on March 3, 1975. They are on display at the Kayoi Whale Museum.[9]

In addition, the Hayakawa Residence of the Hayakawa family, who were the local gentry since the medieval age and the pioneers of classical whaling in Kayoi, was designated as an Important Cultural Property of Japan on February 5, 1974. This is the sole fisherman's residence that was designated as such in Japan. The Hayakawa Residence is made of strong and durable wood; especially thick crossbeams built into the ceiling are impressive. The house also preserved whaling tools, such as a thick hand-knit fishing net. The exceptionally large altar in the altar room of the Hayakawa Residence suggests the depth of belief in Pure Land School Buddhism of the family.[10]

Pure Land School Buddhism

The local residents on Ōmi Island have subscribed to Pure Land School Buddhism to this day. Observing the constant suffering of common people, founder of the Pure Land School Buddhism, Saint Hōnen (1133–1212) propagated a simpler way for them to attain spiritual salvation from poverty, sickness, and other hardships. He taught that people could enter the Pure Land of Paradise (equivalent to heaven) and become Buddhas upon death only by repeatedly and wholeheartedly reciting the chanting word (*nenbutsu*), "*Namu-Amida-butsu*" ("Faith in Amida Buddha") in their daily religious practice, instead of having to learn to read and recite the difficult sutras of Buddhism. The fishermen became parishioners of the local Pure Land School Buddhism Kōgan Temple and piously practiced the *nenbutsu* prayers.[11]

In this context, the local community as a whole solemnly embraced the death of whales, and Kōgan Temple held funeral services and annual memorial services for whales, treating them as if they were human beings. They have retained this practice to this day even after they stopped whaling more than a century ago. In addition, the local whaling guild created monuments and other tokens in order to mourn the deaths of whales. Some of these relics have been preserved. One of them is recognized as a "Nationally Designated Site of Historic Importance" (*kokutei-shiseki*) by the Japanese government, while a second is an Important Cultural Property of Japan, and a third is an Important Tangible Folk Cultural Property of Japan designated by the Japanese government.[12]

Whale Grave on Ōmijima

One of the most important monuments for whales on Ōmi Island is the *Kujira-baka* (Whale Grave) erected in June 1692 at a Buddhist hermitage, called Seigetsu-an, near Kōgan Temple in the Kayoi district on the island. This solitary hermitage was the residence of the retired fifth head priest of Kōgan Temple, Saint Sanyo Shunzui. According to the record of Kōgan Temple, he was born in Kan'ei 5 (1628) and died on November 16, Kyōho 19 (December 10, 1734) at age 106. The Ōmijima Whale Grave is made of granite with a size of 5.6 feet in height × 1.5 feet in width × 0.8 feet in depth. With its two pedestal stones, the gravestone stands almost eight-feet high and magnificently looks over the sea. This gravestone is much taller than ordinary gravestones for human beings.[13]

There is a large inscription on the upper part of the front of the gravestone, "*Namu Amida-butsu*" (Faith in Amida Buddha)—the basic chanting prayer in Pure Land School Buddhism. A paraphrase of the smaller inscription on the lower part reads:

> Your life ended with the death of your mother. We did not intend to catch you. We wanted to release you into the ocean if we could, but we knew that you could not survive without your mother. Please receive our prayers so that your souls may attain the enlightenment of Buddha (see Cover Photograph).[14]

The left side of the gravestone has an inscription of the names of three local whale guild heads: Shitara Magobee, Ikenaga Tōemon, and Hayakawa Gen'emon (they were petitioners who made financial contributions to

Kōgan Temple for making this grave), and the date, "May, Genroku 5" (June 1692 in the Gregorian calendar).[15]

What is unique about this whale grave is that it was created for whale fetuses. As many as seventy-five fetuses were buried in the roughly 180-square-foot lot behind this gravestone. Each body was carefully wrapped with a woven straw mat and then was buried in the ground. While a few graves for whale fetuses have been identified in Japan, such as one in Awa-Katsuyama and one in Ine (see Chaps. 4 and 6), they were made for a single fetus that had been caught on one specific occasion. In contrast, the Ōmijima Whale Grave is a collective graveyard created exclusively for fetuses of whales. This is the only "common cemetery" for whale fetuses that has been preserved. Because of this unique nature, the Ōmijima Whale Grave was recognized as a Nationally Designated Site of Historic Importance by the Japanese government on December 24, 1935.[16]

This whale grave suggests that the fishermen-whalers in Kayoi made a conscious and continuous effort to burying each fetus they found in female whales for nearly two hundred years. The grand size of the whale grave reflects the depth of the sorrow and atonement on the part of the whalers. The former Kayoi Whale Museum director Fujii Fuminori states,

> The fishermen had not intended to kill the fetuses and felt a special sorrow for the fetuses when they found them in the whales they had caught. They knew that the bonds between mother whales and calves were very strong. They felt deep remorse for having killed the pregnant whales for their own living. The whalers built the whale grave on top of a hill, looking over the sea, so that the souls of the unborn calves could see pods of whales swimming in the ocean, and prayed for them.[17]

Memorial Tablet for Whales

Simultaneously with erecting the whale grave, the whaling guild heads commissioned the head priest of Kōgan Temple to make a collective memorial tablet (*ihai*) for whales in June 1692. It is 31 inches in height × 9.0 inches in width × 6.0 inches in depth. It is an exceptionally large memorial tablet. As a measure of comparison, a memorial tablet for a human being is usually about six to eight inches in height. Also, the memorial tablet for whales at Kan'non Temple in Ayukawa is 16 inches in height (see Chap. 3). Moreover, being installed in a gold-painted frame with decorative reliefs and placed on a gold-painted pedestal, the one at Kōgan Temple looks magnificent. On its black-lacquered surface, exactly the same inscriptions as on the Ōmijima Whale Grave are engraved in gold paint.[18]

On the upper part, the basic chanting prayer, "*Namu Amida-butsu*," is engraved in a special sharp writing style that was believed to remove pain and suffering by the power of the sword, while on the lower part, the same dedication as the one on the Ōmijima Whale Grave was engraved in gold paint. On the back, the names of the whale guild heads—Yamada Magosaburō, Ikenaga Tōemon, Shitara Magobee—and other parishioners collectively, who commissioned the memorial tablet, as well as names of Kōgan Temple priests who wrote the inscriptions—retired Saint Nenyo and incumbent head priest Shōyo—were engraved, with the date "May 12, Genroku 5" (June 26, 1692 in the Gregorian calendar, see Photograph 7.1). This memorial tablet for whales was designated as a Tangible Folk Cultural Property by Yamaguchi prefecture on March 22, 1975.[19]

Since 1692, Kōgan Temple has conducted memorial services for whales every spring, after the winter whaling season ended, and whale guild members attended the five-day-long memorial service, wearing *haori-hakama* (formal kimono for men), as Kaneko Misuzu (her penname; real name, Kaneko Teru; April 1903–March 1930) poignantly described in her poem, "Kujira-hōe" ("Memorial Service for Whales," see Chap. 1). Commenting on this poem, Ehime University Department of Agriculture professor Hosokawa Takao states, "Listening to the solemn sound of the bell of Kōgan Temple that signaled the commencement of the memorial service for whales, Kaneko might have 'heard' the cries of the fetuses that had been buried in the Ōmijima Whale Grave. Listening to the solemn sound of the bell that reverberated in the ground, she might have felt the souls of the fetuses."[20]

Kaneko grew up in Senzaki and often visited Kayoi by ferry, where her father Kaneko Shōnosuke (December 1874–February 1906) was born. She stayed at her aunt Miyo's house and attended her father's memorial services (he died in Manchuria before she turned three years old) and summer festivals at the local Sumiyoshi Shrine. Kaneko Shōnosuke was born as the fourth son of Ishizu Sukeshirō and then was adopted by Kaneko Ume (this was a common practice among families who had no male successors). Shōnosuke's great grandmother, Ishizu Man, had become a nun after conducting the seven-year anniversary memorial service of her husband, Ishizu Kōkichi. She was given a Buddhist name, Jōyo-Myōsei, became a disciple of the nineteenth head priest of Kōgan Temple, and died as the head nun of the Shōfuku-an.[21]

Photograph 7.1 Grand memorial tablet for whales, Kōgan Temple, Kayoi, Yamaguchi prefecture, 1692

Thus, Nun Jōyo-myōsei became the spiritual root for the Ishizu family. Kayoi Whale Museum former director Fujii Fuminori, who is related to Kaneko Misuzu (his grandmother, Fujii Chiyo, was Kaneko's cousin), states:

> Kaneko Misuzu visited the grave of Nun Jōyo-Myōsei, as well as the grave of the Ishizu family, during the Buddhist memorial service seasons. She listened to her aunt Maeda Miyo's stories about Nun Jōyo-Myōsei and old fishermen's tales. She also observed the local people's deep benevolence for fish and whales and absorbed it in herself. Her poems that instill gentle empathy with all living things—small and large—were born out of the culture and religion of Kayoi. In essence, Kaneko's poems are an embodiment of the whaling culture of Kayoi.[22]

Death Register for Whales

Even more surprisingly, the whaling guild heads also commissioned the Kōgan Temple head priest to give each one of the whales they had caught a *kaimyō* (posthumous "afterlife" Buddhist name) and to compile a *kako-chō* (*lit.*, "books of the past"; a death register) exclusively for whales to record their afterlife names. The death register for whales is entitled "*Keigei-kako-chō*" ("*kei*" refers to male whales and "*gei*" refers to female whales) and had four volumes. Normally, it was considered that only human beings could attain the enlightenment of Buddha and become a Buddha upon death. Nevertheless, Kōgan Temple head priests gave each one of the whales an afterlife Buddhist name, treating the whales as if they were human beings. Astonishingly, Kōgan Temple gave more than one thousand afterlife names to whales in total, over a period of two hundred years from 1692 (or 1719; Volume One is lost).[23]

Moreover, Kōgan Temple gave male whales the highest-ranking title, "*yo*" (meaning "honorable"), in Pure Land School Buddhism in their afterlife names, as in Saint San-yo Shunzui of Kōgan Temple. This title had only been accorded to people who had received formal Buddhist training at the temple for more than five days. In turn, female whales were given a title, "*myō*" (meaning "graceful or precious"), in their afterlife names. Examples of afterlife names include "Shunyō hakugei" (Honorable Spring White Whale), "Seiyo shō'on" (Honorable Voice of Chanting Sound), and "Kōgetsu myōshō" (Graceful Shining High Moon). These names suggest the depth of respect and remorse felt toward whales on the part of the local fishermen, as well as the enormous financial contributions the whale guilds made to the temple for conferring on whales the highest-ranking posthumous names.[24]

Subsequently, over the course of more than two hundred years, Kōgan Temple head priests accorded more than one thousand afterlife names to whales, including calves. In so doing, they meticulously recorded in the death register for whales, in four volumes, the names of the species of each whale (such as humpback whale and North Pacific right whale), the date and place it was caught, and name of the guild that caught it, in chronological order. Volume One (the oldest book) was lost in a mudslide, but the original book of Volume Two has been preserved. Meanwhile, handwritten copies of Volume Three and Volume Four were transcribed after 1925.[25]

The front and back covers of the original Volume Two are coated with black lacquer for durability and aesthetic beauty. Also, gold paper with the title *Keigei-kako-chō* in black ink is pasted on the front cover. The actual text of the register is in the form of a scroll (continuous paper), which is lined with silver paper for durability and for aesthetic beauty. The register book astonishingly has a total length of more than 22 feet (264.4 inches), and has 34 folds of about 4.0 inches in length. Its width is 9.4 inches. With its gold and silver layout, the registry looks impressive (see Photograph 7.2).[26]

Inside Volume Two are registered 243 afterlife names of whales caught from 1802 to 1842. This indicates that only about six whales were caught per year on average during that period, which in turn suggests that classical whaling was on the decline at this time. In contrast, the existing three volumes together recorded about a thousand afterlife names in total. By commissioning Kōgan Temple to compile special death register exclusively for whales, the fishermen-whalers in Kayoi paid an exceptional tribute to the whales that had sustained their living. No such "Death Register for Whales" was found in other whaling communities in Japan.[27]

There are a few cases in which a local temple gave an afterlife name to a single whale and recorded the name in its death register for human parishioners (such as in Sado, Niigata prefecture; Takayama, Ehime prefecture; and Usuki, Ōita prefecture). Nevertheless, no other case than Kayoi exists where a temple continuously compiled a death register exclusively for whales and gave more than one thousand afterlife names to whales. Given its unique nature, the original death register of Volume Two was designated as a Tangible Folk Cultural Property by Yamaguchi prefecture in March 1975.[28]

Photograph 7.2 Death register for whales, Kōgan Temple, Kayoi, Yamaguchi prefecture, 1692–circa 1908

"Trinity" of Prayers for Whales in Kayoi

The Grand Memorial Tablet for Whales and the Death Register for Whales were originally housed in *Kan'non-dō* (the prayer house of the Goddess of Mercy) in the retirement hermitage, Seigetsu-an of Saint Sanyo Shunzui (the location of the Whale Grave), where he offered prayers every day. Since his death, these artifacts have been preserved at Kōgan Temple. Along with the whale grave, the memorial tablet for whales and the death register for whales constituted a "trinity" of prayers for whales by the local fishermen and are important relics of the culture of mourning the deaths of whales in Kayoi today. By preserving the trinity, Kōgan Temple has conducted the five-day-long memorial services for whales that start on April 29, annually, every year to this day. The local people in Kayoi mourned the deaths of whales as if they mourned the deaths of human beings. It is remarkable that the Kayoi community has continued this practice for more than three centuries, since 1692 (Saint Sanyo Shunzui began it in 1679), and even after it ended whaling operations in 1908.[29]

Buddhist Nunnery, Hōsen-an

In addition, at Hōsen-an, the local Buddhist nunnery in the Ōhibi district in the central part of Ōmi Island, the resident nuns have included whales in their daily prayers. Hōsen-an was originally built as the retirement hermitage for the head priest of Pure Land School Buddhism Saien Temple in the Ōhibi district, Saint Hōgan (1744–1815), who retired in 1792. Saien Temple was restored in 1724 by Saint Sanyo Shunzui of Kōgan Temple, and is known for its strict precepts. Owing to the teachings of the three renowned priests of the temple—Saint Hōgan, Saint Hōshū, and Saint Hōdō—Saien Temple became a mecca for Pure Land School Buddhism in the region and has remained as one of the most renowned temples of Pure Land School Buddhism in Japan.[30]

Among many things Saint Hōgan promoted, he notably organized prayer meetings for children in 1779, which is considered the first "Sunday school" in the world, as recognized by Charles J. Dunn of the University of London. Meanwhile, as local nunneries became fully occupied (there were about seven nunneries that belonged to Saien Temple), the Hōsen-an was turned into a nunnery upon Saint Hōgan's death. As with the case of Buddhist monks, becoming a nun was regarded as an honorable service in the local communities on Ōmi Island, and only the daughters of respectable families could become nuns. For this, the families made considerable financial contributions to the head temple, enough to provide for their daughters for life (as with dowries).[31]

There were eight nunneries in total in Yamaguchi prefecture in 1925, and the Hōsen-an was their head nunnery. A total of forty-seven nuns lived in the eight nunneries, out of which fifteen nuns lived in Hōsen-an. They lived a self-sufficient and frugal life in seclusion under strict precepts and principles of Pure Land School Buddhism. A wooden plaque of the Fourteen Commandments for Nuns is still hung in the living room of Hōsen-an today. The nunnery was off-limits to all men, including the nuns' fathers and their fellow monks. When nuns attended seasonal religious services at their head temple, Saien Temple, they were not allowed to communicate with the monks there, and they had to leave the temple immediately after the service.[32]

Moreover, when they were outside the premises of the nunnery, they had to walk fast and avoid any contacts with local villagers on the streets. They were not allowed to greet or make even eye contact with male villagers. They were not allowed even to have simple conversations with female villagers. Even today, the nuns are pure vegans and eat produce that they grow in their small backyard (they do not eat onions and scallions because they have strong smells). They also grow some flowers, to offer them to Buddha on the altar. There is a small lotus pond, as lotus flowers symbolize the Pure Land of Pardise.[33]

Death Register and Prayers for Whales at Hōsen-an

In 2008, four nuns in their seventies and eighties lived at Hōsen-an, while one nun had been hospitalized. As in the past, their daily life mainly consists of offering prayers—three times a day for a total of six hours—and cleaning the house. According to the teachings of Pure Land School Buddhism, they read aloud each of the afterlife names compiled in their three-volume death register. The death register lists posthumous names of successive Japanese emperors, the nuns' superiors (past head priests of Saien Temple), and their fellow nuns, as well as their parishioners who made regular contributions to the nunnery, according to the day of the month of their death.[34]

For instance, Volume One lists afterlife names of those who died from day 1 to day 10 of each month. Volume Two lists names of those who died from day 11 to day 20 of the month, while Volume Three lists names for those who died from day 21 to day 30 of the month (the lunar calendar has no "day 31"). This filing method enables nuns to read afterlife names on the "monthly memorial day" of the deceased. Just reading aloud the afterlife names listed in the death register takes more than two hours.[35]

The death register of Hōsen-an includes dolphins, whales, and even farm insects. For instance, whales are listed collectively as "*Keigei-gyorui-gun rei*" (Souls of Group of Whales). In addition, the name of Oka Jūrō (July 1870–January 1923, from what is currently Nagato), who is referred to as the "Father of Modern Japanese Whaling," is listed in the book. Oka was concerned with the decline of classical coastal whaling in the Chōshū Kitaura Base, due to Russian whaling operations in the Sea of Japan. As soon as the Meiji government was established in 1868 and it launched Japan's modernization and Westernization, Oka went to Norway to study modern Norwegian whaling methods himself.[36]

Upon returning home, Oka Jūrō founded the Nippon Enyō-Gyōgyō Company (the prototype of present Nippon Suisan), along with Yamada Tōsaku, in Senzaki in July 1899, and began a modern whaling business. He also established a branch office in Shimonoseki, Yamaguchi prefecture. Thus, Senzaki (Nagato) and Shimonoseki became the birthplaces for modern Japanese whaling. Oka also promoted the preservation of many species of whales. The fact that Oka is listed in the death register of Hōsen-an indicates that he made a considerable financial contribution to the nunnery. This in turn suggests that modern Japanese whalers in Nagato also mourned the deaths of whalers, as the classical fishermen-whalers have done.[37]

In August 2012, only one nun, Jikō Shōshin (real name, Hieda Kazuko, b. November 1924) at age eighty-seven, lived in Hōsen-an. She was a native of Ōhibi and became a nun at age thirteen in May 1938. She was first sent to one of the branch nunneries of Hōsen-an, away from Ōmi Island, and served there as an apprentice for thirty years until 1968 when she was sent to Hōsen-an. Recollecting the time of apprenticeship, she states, "When I was lonely, I only thought of going home. When I was happy,. . .well, actually, I was never happy there. I spent countless sleepless nights. I stayed there and persevered only because if I gave up and returned home, my folks would laugh at me."[38]

Former director of the Kayoi Whale Museum Fujii Fuminori wrote in 2014,

> I visited Hōsen-an for the first time in June 2006. I was taken by its peaceful and serene atmosphere so much so that I have visited the nunnery more than 300 times in eight years. In particular, I was not only impressed with the

> beauty of the lotus flowers that bloomed from late June to mid-August, but also with the dedication of Abbess Jikō Shōshin, who has maintained the frugal life of practicing prayers to this day and single-handedly kept the nunnery going for several years.... She will most likely be the last nun at Hōsen-an.[39]

"Discovery" of Stone Statue of Jizō-Bosatsu for the Souls of Whales

In 2004, it was discovered that Hayakawa Genjiemon, in the thirteenth generation of the Hayakawa family (the founder of the whaling guild in Kayoi), commissioned Kōgan Temple to create a stone statue of *Jizō-bosatsu* dedicated specifically to the souls of whales at Kōgan Temple in June 1863. This guardian deity of motherhood and unborn children is believed to ease the suffering of the living and the dead, as a savior from the Six Paths and Three Realms of Desires and Karmic Rebirth. This stone statue for whales is about 2.7 feet in height and stands on a pedestal of about 2.5 feet in height (see Photograph 7.3).[40]

The guardian deity is sitting in the posture of offering a prayer, holding the palms of its hands together. The pedestal has an inscription:

> "*Keigei gyorin gunrei jizō-son*" (Sacred *Jizō-bosatsu* for the Souls of Group of Whales)[41]

American and European whalers had been depleting the Japan Grounds, and whales had stopped coming to Kayoi by that time. The classical Japanese whalers did not realize this because the Tokugawa shogunate government was enforcing a seclusionist *sakoku* policy and contacts with foreign countries and foreigners were banned. Instead, Japanese classical whalers, such as Hayakawa Genjiemon of Kayoi, thought that whales had stopped coming because they had overfished the whales. Hayakawa therefore commissioned Kōgan Temple to make the stone statue of the *Jizō-bosatsu* for whales in order to atone for his sins in the past and to pray for the souls of whales. Thus, about two centuries after Hayakawa Gen'emon had erected the Ōmijima Whale Grave in 1692, his descendent Hayakawa Genjiemon erected the Stone Statue of the Sacred *Jizō-bosatsu* for the Souls of Whales.[42]

Strangely, few people knew about the existence of this statue. Neither the head priests of Kōgan Temple over generations nor the Hayakawa family

Photograph 7.3 Stone statue of guardian deity of unborn children for the souls of whales, Kōgan Temple, Kayoi, Yamaguchi prefecture, 1863

members had recorded anything about it. Then in 2004, Kayoi Whale Museum director Fujii Fuminori discovered it during his field research. He thinks that the existence of this statue had been forgotten probably because it was erected in the graveyard for the head priests of Kōgan Temple and was "hidden" among their graves. Worse, when the statue was temporary removed from the site during a side ditch construction, it was reinstalled in 2001 with its pedestal stone placed backwards. Consequently, the inscription on the front of the pedestal was not visible from the front of the statue and its provenance remained unclear.[43]

Kayoi Whale Song

The Kayoi community has also preserved a whale song the fishermen used to sing upon sailing and/or after the catch. What is unique about this whale song is that singers do not clap their hands loudly, as is typically done in traditional Japanese songs. Instead they only gently rub their hands as if they were praying, in deference to the souls of the whales. The song echoes the Ōmi Island's local people's benevolence for whales and the depth of their belief in Buddhism. The song was designated as one of the "Intangible Folk Cultural Properties" by Nagato city on October 11, 1971. Currently, the Society for the Preservation of the Kayoi Whale Song, led by Hayakawa Yoshikatsu, in the eighteenth generation of the Hayakawa family, performs the song for cultural events.[44]

Kayoi Whale Museum, located next to the Ōmijima Whale Grave, exhibits equipment for classical whaling and various artifacts made of whale bone and baleen, as well as a beautiful wooden statue of *Gyorei kan'non* (the Goddess of Mercy for the Souls of Fish). This statue portrays a scene where the Goddess, wearing a fine kimono in a standing posture, is riding on the back of a mother whale in the sea, with a calf lying at the feet of the goddess. Fujii Fuminori, former director of Kayoi Whale Museum, states, "Local fishermen in Kayoi have not forgotten to appreciate the blessings brought by whaling, and they have prayed for the souls of whales in good seasons as well as bad seasons of whaling. The whaling culture in Kayoi has made Ōmi Island an 'Island of Prayer.'"[45]

In turn, Ehime University professor Hosokawa Takao states that it is astonishing that the Kayoi community has conducted the memorial rites for whales annually, without interruption to this day. The memorial rites for whales, both tangible and intangible, truly make Kayoi the mecca of the culture of mourning the deaths of whales in Japan. The compassion for whales is the backbone of the memorial rites for whales. This is the essence of whaling culture in Japan.[46]

Folklore Legend of Tonomura Family and Whale

Finally, as with several other classical whaling communities in Japan, there is a folklore legend of *nidana* (the Buddhist concept of "chain of causation") about catching a pregnant whale or a whale with a calf in the Nagato region, as a cautionary tale. It is called the "Folklore Legend of the Worship of the Soul of the Whale in Nagato." According to the story, the bay village of Shirakata in Senzaki was in serious financial difficulties because of the poor whale catch. Then, the owner of a local whaling guild by the name of Tonomura had a dream in which the spirit of a whale appeared and stated, "I am passing through off of the bay the next morning with my child. Please do not to kill us. If you grant my wish, I will have myself and my mate caught in your net on our way back."[47]

When he sailed off in the morning, he did find a mother whale with a calf, just as the whale spirit told him in his dream. However, he ignored the plea of the whale spirit and ordered his crewmen to catch the whale. They were elated. Afterward, a series of tragic events befell the Tonomura family, and the prosperity of the family declined. The family eventually disappeared from the Nagato region.[48]

Several studies have erroneously attributed the tradition of the exceptional memorial rites for whales in Nagato to this folklore legend. They argue that the local fishermen treated whales with profound benevolence because of the lesson found in this legend. However, this is not the case. As examined above, the true reason for the tradition of mourning the death of whales in Nagato derives from Saint Sanyo Shunzui of Kōgan Temple in Kaoi. It is true that there is no one by the name of Tonomura in Nagato today. However, Kayoi Whale Museum director Fujii Fuminori discovered six gravestones of the Tonomura family that had been erected at the former site of Entsū Temple in Shirakata, Nagato. From the names and dates inscribed on the gravestones, Fujii managed to track down the descendants of the Tonomura family.[49]

It turned out that Tonomura Shunsuke, in the third generation of the Tonomura family, had been engaged in whaling business in Shirakata in the late nineteenth century. He was also a local mayor and a member of the Yamaguchi Prefecture Assembly, but died at age forty-two. He died in a fire at a local hospital on September 11, 1886, along with his wife and his son.

This appears to be the reason for the association of the Tonomura family with the folklore legend about the whale.[50]

The problem with this association is that the tragedy of the Tonomura family took place in 1886, whereas the memorial rites for whales have been practiced in Kayoi, Nagato since 1692. Thus, the time is wrong by two-hundred years. Moreover, contrary to the legend, Tonomura Shunsuke's two daughters did survive. Also, Tonomura Misao, in the sixth generation of the Tonomura family, used to live in Hagi and currently lives in Yamaguchi, the capital city of Yamaguchi prefecture. Thus, it appears that the cautionary tale about a mother whale with a calf in Kayoi was somehow mixed up with the tragedy of the Tonomura family, which had been engaged in whaling in the late nineteenth century.[51]

Analysis

The culture of mourning the deaths of whales and atoning for the sin of killing them in Kayoi is primarily owed to Kōgan Temple fifth head priest Saint Sanyo Shunzui. He was born Ikenaga Sangō, the second son of the Ikenaga family, head of one of the local whale guilds. The Japanese custom mandated that the eldest son succeed to the family position and profession, and subsequent sons must seek different careers. Having been born to the family of the head of a whaling guild, the young Ikenaga Sangō witnessed the sorrows and remorse of local fishermen who engaged in classical whaling. In this context, the young Ikenaga became a Buddhist monk, which was considered an honorable career by the local community on Ōmi Island.[52]

Upon retiring as head priest of Kōgan Temple in August 1679, Saint Sanyo Shunzui lived in a small hermitage, Seigetsu-an, and devoted his remaining years to praying for the souls of the dead, including those of whales, except for the several years in which he served as the head priest for the restoration of Saien Temple in the Ōhibi district. After the stint at Saien Temple, he returned to his hermitage and built a prayer house there for a statue of the Goddess of Mercy that had been preserved in the locale since medieval times. Because of this history, local people refer to Seigetsu-an as Kan'non-dō (Prayer House for the Goddess of Mercy). Saint Sanyo died there in December 1734, at age 106.[53]

While immersing himself in prayers, Saint Sanyo Shunzui also preached to the local whaling guild members about the need to atone for their deeds and to pray for the souls of whales they had caught. The Ōmijima Whale

Grave, the Memorial Tablet for Whales, and the Death Register for Whales are the products of Saint Sanyo's teachings. That the death register of the nunnery, Hōsen-an in the Ōhibi district, includes whales also reflects the teaching of Saint Sanyo Shunzui. Hōsen-an is a branch nunnery of Saien Temple that Saint Sanyo Shunzui restored. He thus succeeded in making memorial rites for whales a regular practice for fishermen on Ōmi Island despite the fact that only human beings were believed to be able to attain the enlightenment of Buddha and become a Buddha upon death.[54]

While there are a plethora of whale graves and monuments for whales in other whaling communities in Japan, creating a collective grave for fetuses of whales, conferring Buddhist afterlife names on each of the whales caught, and compiling four volumes of a death register exclusively for whales, and thereby regularly conducting memorial services for whales, were unprecedented. This is mostly owed to Saint Sanyo Shunzui.[55]

In addition, being on an isolated island might partly account for the unique aspects of the local whaling culture. The local people had retained the Buddhist concept of benevolence for all living things without much interference from outside world. The town abounds stone statues of *Jizō-bosatsu* wearing red bibs (because they are guardian deities of unborn children due to abortions and miscarriages), including a group of twelve adorable *Jizō* statues in the Hitakiba district. There is even a grave for *Shichi-kan'non* (Seven Goddesses of Mercy). Local residents make offerings of food and flowers to the *Jizō-bosatsu* statues and other statues on the streets every day. This is part of their daily life. The culture of mourning the deaths of whales in Kayoi is an epitome of their benevolence. Kaneko Misuzu nurtured empathy for all the living things in Kayoi and wrote poems that were seen through the eyes of fish and whales, as well as of flowers and trees.[56]

Senzaki

Places elsewhere in Nagato also kept the relics of this whaling culture. Senzaki (incorporated into Nagato City in March 1954) was also a fishing village and is known as one of the ports where Japanese settlers in Manchuria (currently China's Northeast region) were repatriated after the end of World War II in 1945. While the Soviet Army that invaded Manchuria had murdered many of the Japanese settlers, some survivors and orphans reached Huludao port in Liaoning province and sailed back to Japan. As many as 413,961 Japanese were repatriated to Senzaki from China. As the repatriation ships approached Senzaki, the repatriates shed tears looking at the shores and mountains of their homeland. A monument was erected to commemorate this history.[57]

Senzaki had engaged in classical whaling for 235 years, from 1672 to 1907, but the operations in Senzaki were eclipsed by those in Kayoi and few records are available. What is known is that local fishermen and whalers in Senzaki were as religious as their counterparts in Kayoi and that they too had mourned the deaths of whales deeply. This is suggested by the fact that Buddhist temples are a ubiquitous feature of this community and there is one in almost every street. One of them, a Pure Land Buddhism School Gokuraku Temple, has in fact preserved a respectable memorial tablet for whales. A large gold-painted tablet has the inscription, "*Namu Amida-butsu*" (Faith in Amida Buddha), and the same prayer words as the memorial tablet for whales at Kōgan Temple in Kayoi, but in smaller letters.[58]

Another local temple, Hokke Buddhism School or Lotus Flower Buddhism School Fumon Temple, erected a memorial stone dedicated to fish in August 1759 (whales were considered a species of fish at that time). The inscriptions read:

> "*Hokke-kyō ichiji-isseki*" (on the front side); "For the bay villages to prosper, for fish to attain the enlightenment of Buddha" (on the right side); and "...To respect the Dragon God of the Sea for all living things" (on the left side).[59]

"*Hokke-kyō ichiji-isseki*" refers to the prayer of copying down the Lotus Sutra on a tiny pebble—one Chinese character on each—and then burying these pebbles in the grave. In addition, Gokuraku Temple and another Pure Land Buddhism School temple, Enkyū Temple, have conducted annual memorial services for whales in April. Today the two temples take turns in conducting the annual memorial service. Furthermore, local True Pure Land Buddhism (Jōdo-shinshū) temples—Jōgan Temple, Saikaku Temple, and Henshō Temple—have conducted annual memorial services for whales in April. Today, the three temples also take turns in conducting the annual memorial service.[60]

More recently, Saikaku Temple erected a memorial stone dedicated to the souls of fish in November 1965 (Senzaki stopped whaling in 1907). Enkyū Temple followed suit and erected its own cenotaph for fish in November 1976. Further, a venerable local shrine, Yasaka Shrine (formerly Gion Shrine), has preserved a refined picture scroll of *Kujira hokaku-ezu* (Picture of Whaling).[61]

Misumi

The record of classical whaling in Misumi (incorporated into Nagato City in March 2005), located in between Hagi and Nagato, is sparse, but the local history records that True Pure Land School Buddhism School Seifuku Temple in Misumi has conducted memorial services for whales annually, to this day.[62]

Whale Grave and Monuments in Kiwado, Tsuō-Tateishi, and Kawajiri

Moving northwest along the coast to three other fishing communities that had engaged in classical whaling—Kiwado, Tsuō-Tateishi, and Kawajiri (all incorporated into Nagato City in March 2005)—a large stone statue of the Goddess of Mercy in Tsuō-Tateishi in Yuya-chō stands on top of a huge rock and gracefully looks over the Sea of Japan. The local history records that there was a memorial stone for whales at Kaigan Temple in Kiwado, but it had decayed and weathered away.[63]

In turn, Kawajiri, Yuya-chō, has preserved whale graves. The Sea of Japan was rough, making fishing difficult and dangerous in the waters off Kawajiri, especially during the winter. The local fishermen in Kawajiri could not make a living by fishing alone and they travelled far to engage in seasonal physical labor, even risking their lives. In this predicament, the Kawajiri community decided to venture into classical whaling and organized a whaling guild in March 1698. Kawajiri grew to become one of the major stations for the Chōshū Kitaura Classical Whaling Base and prospered until 1908, when it disbanded the whaling guild as the whales had stopped coming off from the Sea of Japan.[64]

During this 210-year period, the fishermen in Kawajiri buried the nasal bone of each whale they caught and erected a gravestone at the whale grave. They also conducted grand memorial services for whales every spring at the end of the whaling season and prayed for the souls of whales according to the Buddhist custom. There is also a record that the whaling guild in Kawajiri made a wooden statue of the Goddess of Mercy for whales in November 1897 and donated it to a local temple. As the original whale gravestone had weathered away in the course of two and a half centuries due to the inclement weather, the Kawajiri Fisheries Cooperative in March 1961 decided to make a new monument and erected a memorial stone there. This giant stone in the shape of a flipper is dedicated to "*Geirin no rei*" (Souls of Whales).[65]

Monument for Whales in Shimonoseki

Moving to the western tip of Yamaguchi prefecture, as noted earlier, Shimonoseki became the birthplace of Japanese modern whaling in 1899 when Oka Jūrō established a branch office of the Nippon Enyō-Gyōgyō Company. Then, in 1922 a fisheries company owner, Nakabe Ikujirō (February 1866–May 1946), bought Tosa Hogei (Tosa Whaling) and began a whaling business in Shimonoseki. Nakabe originally ran his wholesale fish and shipping business in Akashi, Hyōgo prefecture, for the fish market in Osaka, but he moved his operations base to Shimonoseki in 1904 and founded the Hayashikane Company there in 1924.[66]

In 1936, Nakabe Ikujirō began whaling operations in the Antarctic Ocean. The company went through a series of mergers and acquisitions and corresponding name changes, such as Taiyō Hogei (Ocean Whaling) and Taiyō Gyogyō (Ocean Fisheries), and to the current Maruha Nichiro Corporation, a major manufacturing company for processed fish products, such as canned cooked sardines and fish sausages. Nakabe also founded the Nakabe Scientific Research Institute in 1941, which has become the Institute of Cetacean Research today.[67]

The international environment has drastically changed since then, and in 1982 the International Whaling Commission (IWC) adopted its moratorium on commercial whaling, and the Japanese government officially ended commercial whaling in the Antarctica Ocean in 1986. Afterwards, Shimonoseki became a base for "scientific research" whaling. Then, in May 2002, the annual general meeting of the IWC was held in Shimonoseki and a monument for whales was erected in April 2002 in commemoration of the meeting. The monument stands in front of the Shimonoseki City Aquarium, Kaikyō-kan, on the premises of the Shimonoseki Port Park, Arukapōto. A monument made of granite has an inscription, "*Kujira-san arigatō*" (In Appreciation of Whales), along with its translation into English and Latin.[68]

The inscription was written by a local fifth grader, whose handwriting was chosen in a public contest for the creation of the monument. On top of this stone monument stands a bronze statue of a blue whale in a dynamic posture, as if it were diving into the sea. It is a one-tenth size replica of a blue whale. In March 2014, the International Court of Justice (ICJ) ruled that

Japanese whaling operations in the Antarctic Ocean were not for "scientific research" and revoked authorization for whaling there. Following the ICJ decision, the Japanese whaleship Nisshin-maru finished its operations in the Antarctic Ocean and returned home at Shimonoseki port in April 2015.[69]

* * *

Whale Graves and Monuments in Sanyō Region

Not many whale graves and monuments are found in the southern part of the Chūgoku Region—the Sanyō Region—because it is on the coast of the Seto Inland Sea and away from both the Pacific Ocean and the Sea of Japan. Still, some monuments exist in Okayama Prefecture and Hiroshima Prefecture.

Whale Grave in Okayama Prefecture

For instance, there is a whale grave in Ushimado-chō, Oku county (incorporated into Setouchi city in November 2004), located a little to the east of the capital city, Okayama, in Okayama Prefecture. With a calm sea and a moderate climate, jeweled with many islets where olive groves abound, Ushimado is referred to as "Japan's Aegean Sea." In June 1884, a large 98-foot-long whale drifted into the Seto Inland Sea off Ushimado. The local fishermen caught it and towed it to the shore of Odō at the northwestern corner of a major islet of Ushimado, called Maejima. Many people in the neighboring villages flocked to Ushimado and took a ferry to Maejima in order to see the huge whale. Afterwards, the villagers buried the bones of the whale and erected a gravestone for the whale on the shore at Odō. The gravestone is about three-feet tall and has the inscription: "*Taigei kuyō-tō*" (Memorial Stone for Big Whale).[70]

Whale Grave and Monument in Hiroshima Prefecture

Goddess of Mercy for Whales in In'noshima, Onomichi

Moving west to Hiroshima Prefecture, there is a statue of "*Shō-kan'non*" (the Sacred Goddess of Mercy) in Habu-chō, In'noshima (this island was incorporated into Onomichi city in January 2006). The island was the headquarters of the In'noshima Murakami clan, a branch family of the famous Murakami Maritime Force (*suigun*) during the medieval period. The sacred statue is housed in a Rinzai School Zen temple called Bon'on-zan Jigen Temple Taichō-in, which is the No. 26 Temple of the pilgrimage loop of the Thirty-three Setouchi Kan'non Sacred Temples. Local fishermen, who had engaged in passive whaling, donated this statue to the temple in order to pray for the souls of whales. Therefore, local people refer to this statue as "*Kujira kan'non*" (the Goddess of Mercy for Whales). The bronze statue of the Goddess of Mercy for Whales stands next to the main temple building. In addition, there is a large whale *ema* at Hachiman Shrine in In'noshima. Local fishermen donated it to the shrine on February 7 (March 18 in the Gregorian calendar), 1899, in appreciation of the catch of a whale.[71]

Whale Grave in Kure

There is a whale grave in Kure, a little east of the capital city, Hiroshima. The grave stands in a quiet private cemetery on top of a hill, looking over the calm Seto Inland Sea. The gravestone is made of granite and is about 2.0 feet in height × 0.6 feet in width × 0.4 feet in depth. Its front inscription reads: "*Chō-gei-hi*" (Memorial Stone for Whales) in large Chinese characters, while the back inscription reads: "September 1983, Erected by Nakajima Kenzō." It appears that this person had been engaged in modern whaling and created a whale grave in order to pray for the souls of the whales he caught. An extensive internet search for this person could not establish his identity, as a great number of Japanese have the identical name.[72]

Notes

1. "Kujira bunka: Kujira o tomuratta kujira-baka · kujira-zuka nado" (Whale Culture: Whale Graves and Monuments to Mourn the Deaths of Whales), http://www.geocities.co.jp/NatureLand-Sky/3011/kujirahaka.html, accessed November 14, 2015; "An'yō-ji" (An'yō Temple), http://www.eonet.ne.jp/%7Ea-honda/anyojitera/simane2822.htm, accessed December 19, 2015.
2. Yoshihara Tomokichi, "Kujira no haka" (Graves of Whales), in Tanigawa Ken'ichi, ed., *Nihon minzoku bunka shiryō shūsei* (Collection of Documents

on Japanese Folk Culture), Vol. 18, Tokyo: San'ichi-shobō, 1997, 441; "Itae-choshiki Setogashima hogei-zu" (Wooden Plaque Picture of Whaling at Setogashima), http://www.city.hamada.shimane.jp/www/contents/1001000002430/index.html, accessed December 20, 2015.
3. "Hagi-shi" (Hagi City), http://www.city.hagi.lg.jp, accessed July 25, 2012.
4. Ibid.
5. Yoshihara, 448.
6. Kawano (not Kōno) Ryōsuke, *Chōshū · Kitaura hogei no aramashi* (Introduction to Chōshū Northern Bay Classical Whaling), Nagato: Nagato-Ōtsu kujira-shokubunka o keishō suru-kai, 2005, 4–22.
7. Ibid.
8. Tsunoda Fusako, "Ōmi-jima: Nisō no iru chi no hate" (Ōmi Island: The Edge of the World Where Nuns Live), *Tabi*, September 1969, 70–74.
9. Kawano 5, 10–13, 81–82, 86–120; Kobayashi Takako, "Kujira hakubutsukan" (Whale Museum), *Shōhisha*, December 2004, 58.
10. Kawano, 80–81; Kobayashi, 59.
11. Fujii Fuminori, "Ōhibi Amadera Hōsen-an no rekishi: Jikō-Shōshin anshu no inori" (History of Nunnery Hōsen-an in Ōhibi: Prayer of Abbess Jikō-Shōshin), *Kyōdo Bunka Nagato*, No. 26, 2014, 78–87; Kawano, 58.
12. Fujii, 78–87; Kawano, 78.
13. Kawano, 79–80; Fujii Fuminori, email to author, February 5, 2016.
14. Kawano, 79–80.
15. Ibid.
16. Kawano, 78–79.
17. Ibid., 79–80; Fujii Fuminori, author interview, August 9–10, 2012.
18. Kawano, 83–84.
19. Ibid., 78, 83–84.
20. Ibid., 58; Hosokawa Takao, "Nagato kujira-zuka tanbō-ki" (Journal of a Visit to the Whale Grave at Nagato), Part 3, *Yamaguchi Shimbun*, March 1, 2010.
21. Kon'no Tsutomu, *Kaneko Misuzu futatabi* (Revisiting Kaneko Misuzu), Tokyo: Shōgakukan, 2007, pbk., 2011, 43–44, 113–119, 252–254; Kon'no Tsutomu, "Kaneko Misuzu ga nokoshita-mono" (Legacy of Kaneko Mizusu), *Kyōdo Bunka Nagato*, No. 24, 2012, 134–149.
22. Fujii, author interview.
23. Ibid.; Kawano, 83–84.
24. Ibid. (both); Fujii Fuminori, photocopies of the Death Register for Whales, provided to author, August 2012.
25. Kawano, 83–84; Yoshihara, 450.
26. Ibid. (both).
27. Kawano, 78, 83–84; Fujii, photocopies of the Death Register for Whales.
28. Ibid. (both).

29. Kawano, 83–84.
30. Fujii, "Ōhibi Amadera Hōsen-an no rekishi:. . .," 88–89; Iwamoto Chigusa, "Jōdo-shū Amadera: Hōsen-an no seikatsu to shokuji"(Jōdo School Nunnery: Life and Diet of Hōsen-an), *Kyōdo Bunka Nagato*, No. 20, 2008, 82–83.
31. Ibid. (both). Charles J. Dunn is the author of *Everyday Life in Traditional Japan*, North Clarendon, VT: Tuttle Publishing, 1969, pbk., 2008.
32. Iwamoto, 85–90; Tsunoda, 73–74.
33. Ibid. (both).
34. Fujii, "Ōhibi Amadera Hōsen-an no rekishi:...," 90–94; Kumi Kato, "Prayers for the Whales: Spirituality and Ethics of a Former Whaling Community—Intangible Cultural Heritage for Sustainability," *International Journal of Cultural Property* (2007), No. 14, 283–313.
35. Ibid. (both).
36. Ibid. (both); Kawano, 42–52; "Nihon kindai-hogei no chichi Oka Jūrō-shi shikyo" (Father of Japanese Modern Whaling, Mr. Oka Jūrō, Dies), January 7, 1923, http://www.town.abu.lg.jp/sys/topics/detail.php?detailID=394
37. Kawano, 42–52; "Nihon kindai-hogei no chichi Oka Jūrō-shi shikyo."
38. Fujii, "Ōhibi Amadera Hōsen-an no rekishi:...," 91, 93.
39. Ibid., 70.
40. Fujii Fuminori, "141-nen-buri ni nemuri kara sameta o-jizō-sama" (Stone Statue of Guardian Deity that was Awaken after 141 Years), *Nagato kōhō*, December 2004.
41. Ibid.
42. Ibid.; Fujii, author interview.
43. Ibid. (both).
44. Kawano, 78, 84.
45. Fujii, author interview.
46. Hosokawa Takao, "Koshiki-hogei no machi · Kaneko Misuzu no kokyō dearu Nagato-shi o otozure kujira-bunka ni fureru" (Visit to Nagato, Town of Classical Whaling and Birthplace of Kaneko Misuzu, to Observe Whale Culture), *Geiken tsūshin*, No. 447, September 2010, 9–18.
47. Fujii Fuminori, "'Nagato gei-rei sūhai densetsu' o tadasu" (To Question the "Legend about the Whale Spirit in Nagato"), *Kyōdo Bunka Nagato*, No. 25, 2013, 56–64.
48. Ibid.
49. Ibid., 34–64.
50. Ibid.
51. Ibid.; Kon'no (2007), 2011, 213–214, 232.
52. Kawano, 83–84; Fujii, email to author.
53. Ibid. (both).
54. Kawano, 83–84; Fujii, author interview.

55. Ibid. (both).
56. Hosokawa, "Koshiki-hogei no machi..."
57. "Yokuryū-sha hikiage: Senzaki" (Repatriation of Internees: Senzaki), http://www.asahi-net.or.jp/~un3k-mn/hikiage-senzaki.htm, accessed July 11, 2017.
58. Kawano, 5–6, 58; Yoshihara, 450.
59. Kon'no (2007), 217–218; "Gyorei no hi" (Memorial Stones for the Souls of Fish), *Nagato* (city newsletter), December 1, 2004.
60. Ibid. (both); Kawano, 58.
61. Kon'no (2007), 218; "Gyorei no hi"; Kawano, 72–73; Yoshihara, 450.
62. Yoshihara, 450.
63. Fujii Fuminori, "Kujira eno inori" (Prayers for the Whales), lecture at Yuya, Nagato, June 2009; Kawano, 8–9, 34–36, 62–63.
64. Ibid. (both).
65. Ibid. (both).
66. Kawano, 42–52.
67. Ibid.
68. Ibid. 58; "*Kujira kansha-hi*" (Memorial Stone to Thank Whales), July 25, 2003, http://www.asahi-net.or.jp/~rn2h-dimr/ohaka2/12kujira/16cyugoku/simonoseki.html
69. "*Kujira kansha-hi*"; "Chōsa-hogei-sendan ga Shimonoseki nyūkō" (Scientific Research Whaleships Return to Shimonoseki Port), Asahi Shimbun, April 5, 2014.
70. "Taigei kuyō-tō" (Memorial Stone for Big Whale), http://www2a.biglobe.ne.jp/%257emarusan/phokujirakuyoto1.html, accessed December 18, 2015.
71. Yoshihara, 436; "Kujira bunka: Kujira o tomuratta kujira-baka · kujira-zuka nado" (Whale Culture: Whale Graves and Monuments to Mourn the Deaths of Whales), http://www.geocities.co.jp/NatureLand-Sky/3011/kujirahaka.html, accessed November 14, 2015; "Setouchi kan'non reijō" (Kan'non Temples in Setouchi), August 2, 2009, http://sherpaland.net/setouchi-kannon/26/26-top.htm
72. "Chō-gei-hi" (Memorial Stone for Whales/), April 2, 2010, http://www.asahi-net.or.jp/~rn2h-dimr/ohaka2/12kujira/16cyugoku/hiro_cyougei.html

CHAPTER 8

Pilgrimage of Eighty-Eight Temples and Whale Graves in the Shikoku Region

The Shikoku Region is made of four prefectures on Shikoku Island—Kagawa, Tokushima, Kōchi, and Ehime prefecture. Kagawa and Tokushima prefectures, located on the northeastern part of the island, are tucked away from the Pacific Ocean, and have few whale monuments. One exception is a large whale *ema* (*lit.*, "picture horse"; a wooden plaque originally with a picture of a horse, but this case with a picture of a whale) in the Ema-dō (*Ema* House) at Kotohira Shrine in Kagawa prefecture, or the *Konpira-san*, the supreme head shrine of all the other Kotohira Shrines in Japan, which enshrine a guardian god of maritime traffic. In contrast, Kōchi and Ehime prefectures on the southwestern part of the island have numerous whale graves and monuments because of their geographic location.[1]

Facing the Pacific Ocean, the Tosa Classical Whaling Base in Muroto , on the eastern edge of the state of Tosa (current Kōchi prefecture), was one of the four major bases for classical Japanese whaling. Local fishermen sailed out into the ocean, challenging the high waves in search of bonitos and yellowfin tuna, as well as whales. The geographic location also brought tragedies to local residents. In recent history, Muroto is remembered for the Muroto Typhoon, the massive typhoon that landed near Cape Muroto in September 1934, taking a record toll of 3036 people dead or missing and 14,994 injured. Another massive typhoon landed on Cape Muroto in September 1961, with 222 people dead or missing and 4972 injured (the second Muroto Typhoon or Typhoon Nancy).[2]

M. Itoh, *The Japanese Culture of Mourning Whales*,
https://doi.org/10.1007/978-981-10-6671-9_8

Historical Background of Classical Whaling in Kōchi Prefecture

The recorded history of classical whaling in Kōchi prefecture goes back to the late sixteenth century. According to the Tale of Tosa that chronicled the rise and fall of the Chōsokabe clan, the dominant samurai (warriors) of the state of Tosa during the Warring States period, local fishermen caught a whale that was about 32 feet long in Tosa Bay in January 1591 and presented it to their lord, Chōsokabe Motochika (1539–July 1599). Leading a flotilla of several dozen boats and more than a hundred laborers, Chōsokabe in turn brought the "giant fish" to his former enemy turned ruler of Japan, Toyotomi Hideyoshi (March 1537–September 1598), in Osaka Castle. This New Year's gift surprised Imperial Regent Toyotomi and the Osaka folk.[3]

Then around 1624, a local *gōshi* (a man at the lowest rank of the samurai class in the Tokugawa shogunate government), Tada Gorōemon Yoshihira, began classical whaling by the harpoon method in earnest at the bay village of Tsuro at the eastern tip of Tosa Bay (current Muroto), the location of Cape Muroto. However, the harpoon method was dangerous and difficult, and the operation did not work well. About twenty years later, a classical whaler in Owari province (current western part of Aichi prefecture) by the name of Oike Gizaemon sailed to Tosa at the invitation of the highest-ranking official of Tosa province, Nonaka Kenzan (1615–January 1664).[4]

Nonaka promoted industries for the province, including, pottery, beekeeping (apiculture), and whaling, and invited specialists from other provinces. Observing that many whales migrated off Tosa Bay, Oike Gizaemon sent for his family member, Oike Shirōemon, and together they began classical whaling there. Their operations succeeded to a certain degree but did not last long, and they established a new whaling base in Tosa-Saga at the western edge of Tosa Bay (examined below).[5]

In 1664, Tada Gorōemon Yoshihira's son, Tada Kichizaemon Kiyohira (?–1709), was appointed as supervisor of the whaling guild by the local lord. After hearing that Wada Kakuemon Yoriharu in Taiji in Kishū province had perfected in 1675 a new method of whaling, the net method, Tada sailed to Taiji with his younger brother Yoshitsugu and his colleagues in 1681, in order to learn the new method from its inventor in person. In so doing, they disguised themselves as simple crewmen and tried to work at Wada's whaling guild in Taiji, in order to "steal" the net method. During the Edo period, specialized skills and techniques were the exclusive properties of the traditional guilds, by order of the province lord, and giving them away to outsiders was illegal.[6]

Soon the true identity of Tada Kichizaemon Kiyohira was revealed, but Wada Kakuemon Yoriharu did not banish him. On the contrary, Wada was willing to give away his invention to Tada because he felt that he should not monopolize his method and that people should share the blessings of the sea. Nevertheless, his family and senior guild members strongly opposed this, so the issue was brought to the local lord of Kishū province. The second lord of Kishū province Tokugawa Mitsusada was open minded and concurred with Wada. Thus, Wada not only taught his net method to Tada, but also volunteered to loan his own men—ten *hazashi* (blade thrusters) and sixty *kako* (crewmen)—as well as several whaleboats, to Tada and had them accompany Tada on his way back to Tosa in 1683.[7]

As of July 1957, there were about thirty graves designated for "People from Kumano" (the location of Taiji) in the Tada family's temple, Dendō Temple, in Tsuro, Muroto. The oldest graves were made during the Kyōho period (August 1716–June 1735). In the locale, there were also residents who were referred to by their *yagō* (house names) as "Kumano-ya" ("Kumano house"; commoners were not allowed to assume surnames during the Edo period, and they were instead referred to by their house names). This suggests that they were residences of the descendants of the people from Kumano. These facts indicate that while most of Wada Kakuemon Yoriharu's men returned to Taiji, some settled down in Tsuro, Muroto.[8]

In addition, local elders remember the story they heard from their fathers and grandfathers that local fishermen had visited the graves of the "People from Kumano" and offered prayers during a time the whale catch was poor. This is an interesting episode in the history of classical whaling in Japan, telling of the intimate relationship between Taiji and Tosa.[9]

In the bay village of Ukitsu—one of the neighboring villages of Tsuro—a fisherman by the name of Buemon had also been engaged in classical whaling. In 1639, Buemon was given the title of *gōshi* and the surname "Miyaji" by the local lord and was allowed to carry samurai swords, as Miyaji Buemon. Subsequently, in 1687, the whaling guild in Muroto split into the Tsuro Guild led by Tada Kichizaemon Kiyohira and the Ukitsu Guild led by Miyaji Buemon.[10]

Whale Temple in Ukitsu, Muroto

Memorial Stupas and Hanging Bell in Chūdō Temple

Miyaji Buemon, the founder of the Ukitsu Whaling Guild, put up his own money and founded a new temple, Lotus Sutra Myōkoku-ji School Buddhism Chūdō Temple in Ukitsu, in order to atone for the deaths of whales and to pray for their souls so that they would attain the enlightenment of Buddha in heaven. Each time the catch of whales reached thirty-five, Miyaji had the temple erect a memorial stupa for whales and had all the bay villagers attend the memorial service for whales at the temple, in which the head priest read aloud the Lotus Sutra so that the souls of whales could attain the enlightenment of Buddha. These memorial stupas were destroyed by the Muroto Typhoon of 1934, and only some of the pillars of the stupas remain today.[11]

Then in October 1837, one of Miyaji's descendants, Miyaji Sachūi, commissioned a local metal founder, Fujiwara Masatsugu, to create a bronze *bonshō* (a giant hanging bell) in dedication to whales and donated it to Chūdō Temple. Traditionally, giant hanging bells at Buddhist temples were used to summon monks to prayer sessions and also to tell local folks the time of the day. This huge hanging bell is about 3.3 feet tall and stands on a beautiful wooden pedestal curved in the shape of a lotus flower. The inscription on the front of the hanging bell reads: "*Namu myō-hōrengekyō*" (Faith in the Lotus Sutra).[12]

The detailed inscription at the bottom of the pedestal, written by the head priest of Chūdō Temple, Nichigyō, reads:

> Miyaji Sachūi told me, "My family caught about a thousand whales between 1800 and 1837. I just held a memorial service for whales at the temple this spring, but my heart is still not peaceful. My grandfather had intended to donate a hanging bell to the temple, but died before carrying this out. I would like to fulfill my grandfather's wish." I was therefore happy to have the hanging bell made and wrote the inscription on it.[13]

The hanging bell does not exist today, because it was donated to the government's national drive for collecting metals during World War II, in order to manufacture war machinery.[14]

Grand Memorial Tablets for Whales in Chūdō Temple

Then in March 1840, Miyaji Sachūi also commissioned the head priest of Chūdō Temple, Nichigyō, to create a set of three grand memorial tablets (*ihai*) in order to pray for the one thousand whales he had caught. (A record indicates that there is another standard-size memorial tablet for whales at Chūdō Temple, but its provenance is unknown.) The grand memorial tablets are each two feet in height, and three stand next to each other. They are huge memorial tablets, given that the normal size of memorial tablets for human beings is about eight inches. The inscription on the memorial tablet in the center reads: "*Namu myō-hōrengekyō isana kuyō*..." (Faith in the Lotus Sutra as a Memorial for Whales...).[15]

The inscription on the back of this tablet reads:

> Recognizing that he had caught about a thousand whales from 1800 to 1837, Mr. Miyaji had a foundry make a hanging bell for Chūdō Temple and also had me (head priest Nichigyō) read out fifty books of the Lotus Sutra in order to pray for the souls of whales. Mr. Miyaji thereby wishes that whales will be relieved of their sufferings quickly, that they attain the enlightenment of Buddha, and that they gain unlimited blessings in heaven.[16]

Whale Temple in Moto'otsu, Muroto

There is another whale temple called Kongōchō Temple in Moto'otsu, Muroto. Kongōchō Temple, which is commonly referred to as "Nishi Temple" (its official vermillion seal stamp uses the name, Nishi Temple). It was founded by Saint Kūkai (774–April 835) in 807 at the direct request of Emperor Heizei. Kūkai, or Kōbō-Daishi, as he is generally referred to, was born in the State of Sanuki (current Kagawa prefecture) on Shikoku Island. After training at local temples in Shikoku, in Nara (then Japan's capital), and in Tang-dynasty China, Kūkai founded Shingon School Buddhism Kongōbu Temple on Mount Kōya in Wakayama prefecture. With many buildings and statues of Buddha that are designated as National Treasures of Japan, Mount Kōya is on the "Sacred Sites and Pilgrimage Routes in the Kii Mountain Range," which was designated and registered as a UNESCO World Heritage Site in July 2004 (see Chap. 6).[17]

Kongōchō Temple in Muroto is also recognized as a sacred place in Japan, because the young Kūkai went through the stage of austere discipline as a monk there, so that successive Japanese emperors appointed each head priest of Kongōchō Temple in person, up to the tenth generation.

Consequently, Kongōchō Temple became one of the Three Great Temples in Muroto, and lord Chōsokabe Motochika donated an estate to reconstruct the temple that had been damaged over the course of time. Today, Kongōchō Temple houses many Important Cultural Properties of Japan, including Kūkai's own personal belongings, as well as a wooden statue of Amida Buddha and the bronze statue of the Goddess of Mercy.[18]

Kongōchō Temple is Temple No. 26 of the Eighty-eight Temples associated with Kūkai—the twenty-sixth station of the Pilgrimage Loop of the Eighty-eight Sacred Temples on Shikoku (Temple No. 75, Zentsū Temple, is Kūkai's birthplace). At each temple, pilgrims receive an official vermillion stamp of the temple on a scroll and a scroll with eighty-eight stamps serves as a certificate of the completion of the pilgrimage. Although it entails arduous trekking of between 688 miles and 875 miles, depending on the route, this Buddhist pilgrimage has gained popularity among foreign tourists in recent years, as well as among Japanese in general, as a secular spiritual healing experience, rather than a strictly religious pilgrimage.[19]

Kongōchō Temple is referred to as "Whale Temple" by the local residents because there was a memorial tablet for whales at the temple, but its provenance is unknown today. More recently, former modern Japanese whalers erected memorial cenotaphs for whales at the temple. A former executive of Taiyō Gyogō (current Maruha Nichiro Corporation, see Chap. 7) and a native of Muroto, Izui Moriichi (December 1904–July 2002), erected a memorial tower for whales at Kongōchō Temple in April 1960. This tall monument is 7.0 feet in height × 0.6 feet in width × 0.6 feet in depth, and stands on a large three-tiered foundation that is 4.8 feet in height. The inscription reads:

> "*Hogei hassen-tō shōryō kuyō-hi*" (Memorial Cenotaph for the Souls of Eight Thousand Whales Caught)[20]

Izui Moriichi became a whaleship sailor at the age of 15 and a harpooner at age 22. During his lifetime, he caught 8273 whales (5025 in the Antarctic Ocean, 2816 on the coast of Japan, and 432 in the North Pacific Ocean). He chose the harsh life on the sea and once stated, "I sail to the sea again because I prefer the life on the sea where no deceptions and tricks work." Izui also donated a giant bronze hanging bell to Kongōchō Temple dedicated to whales in April 1960. The inscription on the hanging bell reads:

> For the souls of eight thousand whales caught, to attain the three enlightenments of Buddha…[21]

In addition, Izui created a museum for whales called the Geishō-kan at the temple, which houses some of the whaling equipment he used, as well as archival documents on classical whaling in Tosa.[22]

Following in the footstep of Izui Moriichi, another whaler and native of Muroto, Yamashita Takeyata, erected an almost identical memorial tower at Kongōchō Temple in 1967. This tower is 6.7 feet in height × 0.6 feet in width × 0.6 feet in depth, and stands on a large three-tiered pedestal five feet in height. The front inscription is identical to Izui's, but the back inscription reads:

> Born in the fishing community of Muroto, I became a fisherman . . .worked at Taiyō Gyogyō, then at Kyokuyō Hogei at its founding in 1937, and became a principal harpooner in whaling operations in the Antarctic Ocean in 1938. I thank the blessings of the gods and Buddha. . .and pray for the souls of the eight thousand whales caught.[23]

Whale Shrine in Tosa-Saga

In addition, there is a whale shrine in Tosa-Saga, Hata county (current Kuroshio-chō, Hata county) on the western edge of Tosa Bay, in Kōchi prefecture, called Kashima Shrine. Around 1648, a classical whaler in Owari province, Oike Gizaemon, began classical whaling in Tosa-Saga, with his kin, Oike Shirōemon. Their operations were successful and Oike Shirōemon donated a *waniguchi* (a large flattened bell that has a long slit at the bottom; it looks like a pair of cymbals put together) to Kashima Shrine in April 1652 in dedication to the Ebisu God. *Waniguchi* literally means the "mouth of a shark" (*wani* referred to sharks back then, but it refers to crocodiles today) because the long slit looks like the mouth of a shark as seen from below. The bell is tied to a thick rope that hangs from the beam in the front structure of a shrine or temple. Visitors ring the bell before offering prayers.[24]

The inscription on the center of the bell reads:

> Subscription to the Lotus Sutra so that whales could attain the enlightenment of Buddha quickly

Inscriptions around the edge of the bell read:

> "Faith in Amida Buddha" and "Faith in Abundant Treasures Buddha"[25]

Oike Shirōemon also donated similar *waniguchi* to other sacred places in the locale, such as Chika Temple and Nishi Temple (a different temple from the namesake in Muroto) in Tosa-Saga, as well as Hachiōji Gongen in Ukitsu (Gongen is a local guardian god, a manifestation of Buddha in the form of an indigenous god in Japan, as a result of the syncretism of Tendai School Buddhism and San'nō Shintoism). The *waniguchi* at Hachiōji Gongen has the same inscription as that at Kashima Shrine in Tosa-Saga. Oike also donated a giant drum, five feet in diameter, to Kongōchō Temple (Nishi Temple) in Muroto that was founded by Saint Kūkai.[26]

Jizō-bosatsu Stone Statue for Whales in Kubotsu, Tosa-Shimizu

Going further to the southern edge of Shikoku Island, there is a unique memorial statue for whales in Kubotsu, Tosa-Shimizu. The city is the hometown of Nakahama Manjirō (January 1827–November1898), the first Japanese to engage in modern whaling on an American whaleship (see Chap. 2). An impressive statue of Nakahama John Manjirō ("John Mung") looks over the Pacific Ocean at Cape Ashizuri at the southernmost tip of Shikoku Island. Cape Ashizuri is one of the highlights of the Ashizuri-Uwakai National Park. Along with the Sea of Uwa in Ehime prefecture, this national park offers spectacular views of the ocean. In 1812, Okumiya Masataka, head of the Tsuro Whaling Guild, commissioned the head priest of Shingon School Kaizō-in Temple, located in nearby Cape Ashizuri, to create a stone statue of *Jizō-bosatsu* (Guardian Deity of Motherhood and Unborn Children) for whales.[27]

The stone statue of a Buddha-like figure is about three feet in height and sits on a three-foot-tall pedestal. The inscription on the pedestal reads:

> "Henro-michi, Tame kujira-kuyō nari" (On the Shikoku Pilgrimage route, For the Sake of a Memorial for Whales), Kaizō-in Jū-hōin Yūten[28]

That this statue is of a Guardian Deity of Motherhood and Unborn Children suggests that the whaling guild members wanted to atone for the deaths of fetuses that were found in the whales they caught, as well as of the mother whales. Kaizō-in Temple is located on the route of the Eighty-eight Temple Shikoku Pilgrimage, and many pilgrims visit the temple on their way, but it is not one of the eighty-eight temples.[29]

Many species of whales, such as humpback whales and North Pacific right whales, stopped migrating off Tosa Bay, and classical coastal whaling ended in Tosa at the beginning of the twentieth century. Currently, smaller species of whales, such as Baird's beaked whales and short-finned pilot whales, pass through off Cape Muroto during their spring migration, and Tosa Bay has become a tourist spot for whale watching today.[30]

* * *

Whale Graves and Monuments in Ehime Prefecture

Moving to the western coast on Shikoku Island, where the Bungo Channel and the Hōyo Strait separates Ehime prefecture on Shikoku Island from Ōita prefecture on Kyūshū Island, there were no whaling bases in Iyo province (current Ehime prefecture) because it is away from the Pacific Ocean. However, surprisingly, there are about twenty whale graves and several monuments for whales in Iyo. Unlike the classical whalers in Tosa (Kōchi prefecture), local people in Iyo did not engage in classical whaling, but occasionally caught whales that had washed ashore in their bay communities, a process referred to as "passive whaling." Today, scientific studies have discovered that a smaller species of whale called *katsuo kujira* (Eden's whale)—which turned out to be a different species from *nitari kujira* (tropical whale)—used to swim into the bay area of Iyo from Tosa Bay and elsewhere. Although local fishermen in Iyo caught a far fewer number of whales, they still mourned the deaths of beached whales and created graves for them.[31]

Two Whale Graves in Ainan-chō

On the southwestern coast of Ehime prefecture, there are two whale graves in Nishiumi-chō (incorporated into a newly created town, Ainan-chō, in October 2004). The older one was made in April 1882, and the other in September 1900. On April 29, 1882, a whale of more 42 feet in length jumped and was stranded in a bay of Uchidomari, Nishiumi-chō, and the local fishermen caught it. It happened on the inauguration day of a new building of the local elementary school. The school principal regarded the catching of the whale as a blessing of the gods, because they had worshiped the gods and also were dedicated to education. The community created a grave for the whale at a local temple, Sairin Temple, and erected a gravestone. It is 3.5 feet in height × 1.1 feet in width × 0.7 feet in depth, standing on a pedestal. The inscription was written by the school principal, which reads:

"*Tobiagari-kujira no tsuka*" (Grave for Jumping Whale).[32]

On September 20, 1900, another whale jumped and was stranded in a bay of Ōhama, Nishiumi-chō (current Ainan-chō). The community made a grave for the whale at Sairin Temple and erected a gravestone. It is 2.3 feet in height × 0.8 feet in width × 0.6 feet in depth, standing on a pedestal. The inscription reads:

"*Kujira no tsuka*" (Grave of Whale).[33]

In 1995 the two whale graves at Sairin Temple were moved to the roadside of the temple and the remains of the whales were reinterred there when the temple built a nursery school on the premises. Thus, the two graves stand side by side on the roadside today. On their right stands a signpost stating that these two gravestones are cultural properties of the town.[34]

In addition, there is an interesting festival concerning a whale in neighboring Kashiwazaki, Uchiumi-chō, called "*Ryō-dome no matsuri*" (No-fishing-day Festival). On September 6, 1718, a local fisherman in Kashiwazaki was fishing off the Yura peninsula, the southwestern tip of Shikoku Island, facing Kyūshū Island across the Bungo Channel. While fishing, his ship was about to be overturned by a whale. He prayed to Yura Gongen (a local guardian god; Yura Gongen Shrine still exists today on the peninsula) by stating, "If you save my life, I pledge that I will have the folks of my village, as well as their descendants for generations to come, never go fishing on this day." His prayer was answered and since then fishermen in Kashiwazaki have not fished on September 6, and have celebrated the day instead. The community has retained this tradition to this day.[35]

Three Whale Graves in Uwajima

Moving up the western coast of Ehime prefecture, there are seven whale graves in Uwajima, a major city in southern Ehime. Uwajima used to be an island. Out of the seven graves, five are in Yusu village (incorporated into

Uwaumi village, Uwajima, in April 1958) at the tip of the Komobuchi peninsula (which is extended from the Miura peninsula) protruding into the Sea of Uwa. The area is part of the Ashizuri-Uwakai National Park, with a beautiful "ria coast" (a coastal inlet formed by the partial submergence of a river valley) patched with islands and rock formations.[36]

Whale Grave at Miji-no-shima, Yusu

Local fishermen in Yusu found a whale caught in a rock reef near an uninhabited island in the Bungo Channel, called Miji-no-shima. The island is located off Tsunoura port on the southern side of Uwajima Bay, and faces Akehama (not Akihama)-chō-Takayama, in the city of Seiyo, on the northern side of Uwajima Bay (examined below). The fishermen were not able to free the whale from the rock reef, and the whale died. They felt sorry for the whale and erected a gravestone for it on the island, and prayed for its soul. The memorial stone is about 1.7 feet in height × 0.4 feet in height × 0.3 feet in depth. It is made of aqueous rock (a kind of sedimentary rock) and part of the inscription has weathered away and is illegible, but the remainder reads:

> "*Tobi-kujira-tō*, [ ] *ka* 3 [ ] 4-*gatsu*" (Memorial Stone for Jumping Whale, April...)[37]

The description suggests that the whale was jumping hard in order to escape from the rock reef. The three illegible characters denote part of the year the stone was erected. They suggest that this monument was erected in either April of Bunka 3 (May 18, 1806–June 16, 1806) or April of Kōka 3 (April 26, 1846–May 24, 1846), because there are only two periods during the Edo era that end with "ka."[38]

Whale Grave at Futanarabi-jima, Yusu

The local fishermen in Yusu also created a grave for a whale in 1834. This grave stands on another small uninhabited island in the Bungo Channel, called Futanarabi-jima (*lit.*, "two islands side by side"), which is actually made of two islands as the name indicates. The gravestone is about 2.6 feet in height × 0.8 feet in width × 0.7 feet in depth. The inscription reads:

"*Sangai banrei-tō, Geigai-rei no tsuka, Tame gyoji-an'ei*" (Memorial Stone for All the Souls in the Three Worlds, Grave for the Soul of Whales, For safety and prosperity of fisheries)[39]

Whale Grave at Amayadori-jima, Yusu

Recently, in 1975, local fishermen in Yusu again created a grave for a whale on a small uninhabited island in the Bungo Channel, called Amayadori-jima. A whale was entangled in pearl cultivation rafts near Amayadori-jima. The local fishermen working in the pearl cultivation business could not free the whale from the rafts and the whale died on May 4, 1975. Out of benevolence for the whale, the local pearl cultivation cooperative union made a grave for the whale and erected a gravestone so that its soul could rest in peace in heaven. The memorial stone is about 3.6 feet in height × 0.7 feet in width × 0.6 feet in depth.[40]

Memorial Stone for Whales at Koyanoura, Yusu

Then in 1977, Hotta Minoru, a local fisheries industry entrepreneur in Yusu, erected a collective memorial stone for whales near Shinmei Shrine, in the bay called Koyanoura, in Yusu, Uwajima. Currently, Hotta is president of the Hotta Minoru Hamachi (yellowtail) Farm in Yusu.[41]

Whale Grave at Banshō, Yusu

Hotta Minoru also created a grave for a whale that died in Banshō in Yusu and erected a memorial stone in 1986. The grave is called "*Banshō no kujira-zuka*" (Whale Grave at Banshō) and its inscription reads:

"*Kujira no kuyō-tō*" (Memorial Stone for Whale), Died on October 15, 1986, Erected by Hotta Minoru[42]

Whale Grave at Takasuke, Komobuchi

At the western tip of Komobuchi peninsula lies a bay called Takasuke in Komobuchi village (incorporated into Uwaumi village, Uwajima, in April 1958). Local history records that on February 13, 1892, a whale washed ashore, having been chased and injured by orcas (killer whales). The villagers caught the whale and bought back a mountain with the profit from the sale of the whale. Then, they conducted a funeral service for the whale at the local Mankichi Temple and prayed for its soul. They also created a grave for the whale at the temple and erected a gravestone. The memorial stone is 1.7 feet in height × 0.9 feet in width × 0.6 feet in depth. The villagers even commissioned the temple head priest to accord a *kaimyō* (a Buddhist afterlife name) to the whale. The afterlife name given to the whale was "*Kaiyo geion* (or *geiin*)," which roughly means "Honorable Follower of the Buddha in the Sea with the Sound of a Whale." The head priest then recorded the afterlife name in the *kako-chō* (a death register), which has been preserved to this day.[43]

Whale Grave at Mizutani

On December 25, 1948, fishermen in Mizutani, Uwajima, found a sei whale (*iwashi kujira*) caught between rocks. After catching the whale, they found that it carried a fetus. They sold the mother whale, but buried the fetus where the mother whale was found. They also erected a wooden memorial grave marker on the rocks, with the inscription:

> "*Kujira no tsuka*, December 26, Shōwa 23" (Grave of Whale, December 6, 1948) The grave marker has weathered away and is lost now.[44]

Whale Graves in Akehama-chō, Seiyo

Hayanote Whale Grave in Takayama, Akehama-chō

Moving up the western coast of Ehime prefecture a little further from Uwajima, there are three whale graves in Akehama-chō in the newly created city Seiyo (*lit.*, "western Iyo"). All three whale graves were designated as Tangible Folk Cultural Properties of Akehama-chō in October 1974. Of the three, the oldest in Akehama-chō is located in the district of Takayama, which is the seat of the city government of Seiyo. It is called "*Hayanote no kujira-zuka*" (Hayanote Whale Grave), named after the exact location, and has a compelling history to tell. On July 23, 1837, a pilot whale washed ashore at the fishing community of Akehama-Takayama (pilot whales are often known to wash ashore chasing a school of squid that spawn near the coast). The local villagers had been suffering from the Great Famine of Tenpō (named after the imperial reign era), which began in 1833 and lasted until 1839.[45]

The local head fisherman immediately reported this to the local province government and obtained permission to catch it. The fishermen caught the whale and the villagers gratefully consumed it (whales were considered a species of fish and eating fish was condoned in Japanese society). Thus the whale saved their lives. Also, they did not waste any of its parts. For instance, they extracted oil from the bones and used it as an insect repellant for farming and housing. Then, the villagers buried some of the bones and made a grave for the whale at the shore of Hayanote. They also conducted a funeral service for the whale at the local Kongō Temple.[46]

However, this was not all that the community did for the whale. The seventh lord of Uwajima province, Date Munetada Shunzan (October 1792 or earlier–November 1889) had the head priest of his own temple, Ryūge-zan Tōkaku-ji, accord the whale a *kaimyō* (an afterlife name), treating the whale as if it were a human being. Lord Date even personally handwrote the afterlife name of the whale for the purpose of engraving the name on memorial objects. The afterlife name given to the whale was "*Rin'ō-inden hōkkai-zenka* (or *zenki*) *daikoshi*" (Great Follower of Amida Buddha in the Kingdom of Fish). This is an exceptionally respectable name even for a human being: It contains two of the highest titles, "*inden*" and "*daikoshi*," which were only given to the lord of a province. In comparison, the highest-ranking official of a province was given lower titles, "*in*" and "*koshi*."[47]

At the end of 1838, Lord Date Munetada Shunzan sent an officer to Akehama-Takayama village to deliver his handwritten afterlife name for the whale and 100 moku (a monetary unit for a silver coin) in order to create a *shidō* (a small shrine as a grave) for the whale at the three-year anniversary memorial service for the whale (according to the old counting system. A three-year anniversary memorial service takes places on the two-year anniversary day). Then, the local Kongō Temple head priest in Akehama-Takayama made an *ihai* (a memorial tablet) for the whale and engraved the afterlife name of the whale, handwritten by Lord Date, on it. The head priest also recorded the afterlife name of the whale, along with the provenance of the whale, in the temple's *kako-chō* (death register). This register has been preserved to this today. The temple also erected a gravestone for the whale, which is about 3.6 feet in height × 0.9 feet in width × 0.6 feet in depth. The inscription reads:

> "*Rin'ō-inden hōkkai-zenka* (or *zenki*) *daikoshi*" (Great Follower of Amida Buddha in the Kingdom of Fish), "June 21, Tenpō 8" (July 23, 1837, the day the whale died)[48]

Since then the villagers have worshiped the whale and referred to it as *kujira* (whale)-*sama*, with the honorific suffix. Kongō Temple has conducted memorial services for the whale annually to this day and the local people visit the Hayanote Whale Grave and pay tribute to the whale that saved the lives of their ancestors. The community also performed a whale dance in commemoration of the 150-year anniversary. The gravestone used to stand on the shore at the edge of National Route 378 (near current Akehama Seaside Sun Park) and was housed in a wooden framed glass structure (that looked like a large telephone booth) in order to protect it from the exposure to the weather.[49]

Recently (before May 2013), the gravestone was moved, along with the wooden structure, to the top of a rock formation nearby, due to a road repair, and also in order to protect the monument from high waves. Also, near the shore of Akehama-Takayama, there is an Ebisu Shrine (that worships whales) in the compounds of Kamo Shrine. This venerable shrine houses many spectacular wooden buildings decorated with "open work wood carving" (*sukashi-bori*) objects, such as *hōō* ("Chinese phoenix"; the sacred bird in Chinese mythology, "fenghuang" in Chinese pinyin) and dragons, of first-class techniques.[50]

Kusunoura Whale Grave, Akehama-chō

Another whale grave is located on a bay called Kusunoura on National Route 378, between the Tawarazu district and the Karie district of Akehama-chō, east of the Takayama district. The inscription on the gravestone reads:

> "*Meiji 3, Ah, tairin-geishi haka, kumijū ken'nō*" (1870, Ah, Grave for a Calf of a Large Whale, erected by the community)[51]

The engraving suggests that the grave was made for the calf of a whale that was caught at that time. The use of the exclamatory word "Ah" expresses the deep sorrow of the villagers for catching the calf. The size of the gravestone is about 3.8 feet in height × 0.8 feet in width × 0.6 feet in depth. As with the Hayanote Whale Grave, this memorial stone is also housed in a wooden-framed glass structure. This monument used to stand at the shore, but it was moved to a higher location on the roadside of National Route 378 in 1981 for protection from high waves.[52]

Whale Grave at Miyanotsu, Akehama-chō

The third whale grave in Akehama-chō is located in the Miyanotsu district, west of the Takayama-district, and is called "*Komochi-iwa no kujira-zuka*" (Whale Grave at Komochi Rock). On March 13, 1907, local fishermen caught a 42-foot-long North Pacific right whale that had swam into the bay called Miyanotsu, along with a sailboat, as if it were accompanying the boat that had sailed from the Kyūshū region. A local ship carpenter, who was repairing a sailboat from Osaka, caught the whale with a net, and tied it to a pine tree on Kuchinao Rock, near Komochi Rock. Then, all the villagers in Miyanotsu took part in seizing the whale, but they did not know how to eat whale meat or extract oil from its carcass (the era had changed to the Meiji period from the Edo period and the mores of the local residents had changed), so they shipped the whale to Osaka and elsewhere without gaining a profit.[53]

The villagers still felt sorry for the whale and buried some of the bones at Ebisu Shrine, and erected a gravestone there. The memorial stone for the whale stands about 2.3 feet in height × 0.7 feet in width × 0.7 feet in depth, without its pedestal. The inscription reads:

> "*Tomurau taigyo no rei, March 13, Meiji 40*" (To Mourn the Soul of the Big Fish, March 13, 1907)[54]

The community moved the gravestone to a rock reef near where the whale was found. It was later moved several times, each time that land reclamation was made, and finally to its current location at Komochi Rock (hence, the name, Whale Grave at Komochi Rock). The area was surrounded by the sea in 1907, but Komochi Rock now stands by the shore next to National Route 378.[55]

Whale Graves in Ikata-chō, Nishi-Uwa County

Moving further to the north to the Sata-Misaki peninsula, remarkably, there are five whale graves in Ikata-chō, Nishi-Uwa county, alone. The Hōyo Strait rapidly runs through Cape Sata at the tip of the peninsula toward the Seto Inland Sea (Iyo-nada Sea specifically), and whales used to pass through the narrow strait from the Pacific Ocean into the Seto Inland Sea via the Bungo Channel during the Edo period. Today, the town has the Ikata Nuclear Power Plant on the northern side of the peninsula. This power plant is the only nuclear power plant in Japan that was built on the coast of the inland sea, as opposed to the coast of the outer seas, such as the Pacific Ocean or the Sea of Japan. The Ikata Nuclear Power Plant had supplied about 20 percent of Shikoku Island's energy needs, but operation has been suspended since January 2012 as the result of the accidents at the Fukushima Daiichi Power Station in Fukushima prefecture in March 2011.[56]

WHALE GRAVES AT MITSUKUE, IKATA-CHŌ

There is a whale grave in Mitsukue. In 1809, a whale entered into the Bay of Mitsukue in Seto-chō (which was incorporated into Ikata-chō, Nishi-Uwa county, in April 2005) on the northern side of Sata-Misaki peninsula facing the Seto Inland Sea, chasing a large school of sardines (called "*iriko*" locally). Being unable to turn its body toward the ocean, the whale ran aground and was caught by local fishermen. They were surprised to find that the big fish's blood was warm and they mourned its death. They made a grave for the whale on the sand spit in the bay, which the local people called Suga-no-hama, and erected a gravestone in order to pray for its soul. The memorial stone is about 3.4 feet in height × 0.9 feet in width × 0.6 feet in depth. The inscription reads:

> "*Myōgei no kurai*" (The Rank of Graceful Whale), erected by Mitsukue community[57]

This inscription suggests that the villagers conducted an elaborate funeral service for the whale at a local temple and commissioned the head priest to accord a *kaimyō* (afterlife name) to the whale. Due to exposure to the inclement weather at the sea, the top part of the gravestone has crumbled and the height has shortened to about 1.1 feet today. Meanwhile, Suga-no-hama has become Suga Park and thus the gravestone stands inside the park.[58]

WHALE GRAVE AT KOJIMA, IKATA-CHŌ

There is another whale grave in Kojima, Ikata-chō, along the Sata-Misaki peninsula. In 1879, a whale ran aground on the rock reef and died. The local fishermen caught it and the village folks shared the meat. In order to pray for the whale's soul, the community buried some of the whale's bones and erected a gravestone. It is 6.2 feet in height × 2.0 feet in width × 1.0 foot in depth. This memorial tower is made of the famous indigenous rock of the locale, called *Iyo aoishi* (*lit.*, "blue rock of Iyo"; a kind of chlorite schist), a beautiful green stone marbled by white lines. The inscription on the monument reads:

> "[ ]-*kai gyorin zengei no kurai*" (The Rank of Good Whale of [] Sea).[59]

As with the whale grave at Mitsukue, this inscription suggests that the villagers commissioned the local temple to accord a *kaimyō* (a Buddhist afterlife name) to the whale, as well as to conduct a funeral service for it. The local people refer to the grave as "*Kojima no kujira-zuka*" (Whale Grave at Kojima).[60]

Whale Grave at Kawanohama, Ikata-chō

Going west further down the Sata-Misaki peninsula toward its tip, one arrives at a bay area called at Kawanohama, Ikata-chō. It is located almost in the middle of the Sata-Misaki peninsula on its southern side, facing the Sea of Uwa. On February 1, 1887, a 33-foot-long whale swam into the Sea of Uwa and then washed ashore in Kawanohama. Local fishermen caught it, and buried some of its bones. Then they erected a gravestone for the whale there: 6.3 feet in height × 2.5 feet in width × 0.6 feet in depth. This impressive memorial tower is made of the beautiful *Iyo aoishi* (chlorite schist) and the local people refer to it as "*Kawanohama no kujira-zuka*" (Whale Grave at Kawanohama).[61]

The inscription reads:

> "*Daijō-myōten ichiji-isseki geirei-tō*" (Memorial Tower for the Soul of the Whale with offering of the prayer of one character-on-one-stone of Mahayana Buddhism Sutra), "*Tatematsuru tenka-taihei*" (In dedication to the peace of the world), "*Osameru urajū anzen*" (Donating for the sake of the safety of this bay), erected by Mizuno Tōkichi[62]

Mizuno was most likely the head of the local fisheries cooperative union. This engraving indicates that Mizuno commissioned the local temple head priest to offer, as part of the funeral service for the whale, *ichiji-isseki*, the special prayer of copying down the sutra of Mahayana Buddhism on a tiny pebble—one Chinese character per pebble—and then burying these pebbles around the grave. This elaborate memorial service in turn suggests that the fishermen might have feared punishments from the gods for catching the whale, as well as having felt genuine sympathy for it.[63]

Whale Grave at Natori, Ikata-chō

Going further west from Kawanohama toward the tip of the Sata-Misaki peninsula, on the southern side one reaches another bay community, Natori, Ikata-chō. The whale grave there was made for a sei whale (*iwashi kujira*) around 1929. It is about 1.4 feet in height × 0.6 feet in width × 0.4 feet in depth and has the inscription, "*Hogei kuyō-tō*" (Memorial Stone for Whaling). The local people refer to it as "*Natori no kujira-zuka*" (Whale Grave at Natori).[64]

Whale Grave at Kushi, Ikata-chō

Going down the Sata-Misaki peninsula further, almost at the western tip, one arrives at another bay community, Kushi, on its northern side, facing the Seto Inland Sea. Around 1882, a whale washed ashore in Kushi, and the local fishermen buried some of its bones and erected a gravestone. This respectable memorial tower is made of *Iyo aoishi* and stands 8.0 feet tall and 2.2 feet in width and 0.4 feet in depth. There used to be two memorial stones for the whale, but they weathered away in time, and the community erected a new memorial tower. The local people refer to it as "*Kushi no kujira-zuka*" (Whale Grave at Kushi) or "*Taigyo-sama*" (Revered Big Fish).[65]

Whale Grave in Iyo

Moving north on the coast of the Iyo-nada (part of the Seto Inland Sea), one arrives at Iyo, which is located six miles south of Matsuyama, the capital city of Ehime prefecture. On March 20, 1910, a sperm whale (or a gray whale; the species is uncertain) washed ashore in Kamihama, Iyo, and local fishermen caught it. The villagers buried some of the bones on a hill and made a grave for it. Then, they moved the grave to the local Minato Shrine in Minato-chō near the shore. Strangely, after the fishermen had caught the whale, they continuously suffered a poor catch of fish. They felt that this was a punishment of the gods for catching the whale and erected a tall memorial tower for the whale in order to ease the whale's soul. The monument is

5.0 feet in height × 1.0 foot in width × 0.9 feet in depth. With its pedestal, the tower stands 6.7 feet tall. The inscription reads:

> "*Koku-gei ichiji-isseki-tō*" (Memorial Tower for Overcoming the Whale with offering of the prayer of one-character-on-one-stone), erected by the County Fisheries Cooperative Union.[66]

This engraving indicates that the fishermen commissioned the local temple to offer the elaborate prayer of *ichiji-isseki*, copying down the Mahayana Buddhism sutra one-character-on-one-stone.[67]

As if this monument was not enough, the fishermen also erected another memorial tower for the same whale in the Ozaki district of Iyo (current Ozaki Park), near the shore, a half-mile south of Minato Shrine. It has the same inscription as the first one and the same date. The only difference is that this one was erected by Ōmoto Jōkichi, who was most likely the head of the fisheries cooperative union. The size of this memorial tower is 6.2 feet in height × 0.7 feet in width × 0.5 feet in depth. That the fishermen erected two respectable monuments for the whale, as well as offered the elaborate prayer of *ichiji-isseki* for the whale's soul, indicates that the fishermen feared punishment from the gods for catching the whale and prayed for forgiveness.[68]

Memorial Stones for Fish in Iyo

In addition, there is an interesting group of memorial stones for fish in Iyo-Kaminada, Futami-chō, Iyo. Going south on National Route 378 from Iyo, one arrives at a village called Koami. There, a total of twelve memorial stones (two of which have already crumbled for the most part) stand side by side on the roadside of the Old National Route 378 (the new Route 378 runs right below the old one) that runs along the coast. One of the memorial stones is for sardines and the engraving reads:

> "*Namu Amida-butsu tatematsuru kuyō tame tairyo iwashi* [ ]" (Faith in Amida Buddha in dedication to the souls of sardines caught and for a good catch in the future).[69]

The date of erection was October 16, 1889. Another memorial stone is for octopi, while the oldest one was dated sometime in the Kyōho period (August 1716–June 1736). At the center of the twelve memorial stones, a statue of *Shō Kan'non-bosatsu* (the Sacred Goddess of Mercy) is housed in a small shrine, which is flanked by stone statues of a set of the Six *Jizō-bosatsu* (Guardian Deities of Motherhood and Unborn Children), three on each side. In turn, three statues of the Six *Jizo-bōsatsu* are flanked by the twelve memorial stones for fish, two on the left and ten on the right side. These statues and memorial stones suggest the local fishermen were saddened by having to kill fish and other marine animals in order to sustain their own livelihood. Therefore, they sincerely prayed for the souls of the marine animals so that they would attain the enlightenment of Amida Buddha.[70]

Kawanoe-chō, Shikou-Chūō

Moving east from Iyo to the northeastern edge of Ehime prefecture, there lies the fishing community of Kawanoe (this city was abolished and incorporated into a newly created city called Shikou-Chūō in April 2004, and became Kawanoe-chō). In 1864, the local fishermen in Kawanoe caught two whales successively, one on February 21 and another on February 27. As a memorial for the whales, the villagers made a large whale *ema* out of a whale bone and donated it to the local Hachiman Shrine.[71]

* * *

One might wonder why there are so many whale graves in Iyo, current Ehime prefecture—more than in Kōchi prefecture, the site of the Tosa Classical Whaling Base—where there were no classical whaling bases. Fishermen's villages in both Iyo and Tosa were religious communities, constituting the route of the Eighty-eight Temple Shikoku Pilgrimage. Both were superstitious, as was the rest of Japan. The local people on Shikoku Island had a hospitable spirit so that even today they treat random pilgrims, who are recognized as such by the white pilgrimage kimono they wear, with a cup of green tea or a sweet. What is the difference then?

Ehime University Department of Agriculture professor Hosokawa Takao points out that fishing communities in Tosa Bay were confronted by the high seas of the Pacific Ocean, whereas those in Iyo were tucked into the inland sea of Bungo Channel and away from the ocean. Iyo was blessed with a moderate climate where mikan (tangerine oranges) grow (Ehime is one of the major mikan growers in Japan), and the villagers in the calm bay communities in Iyo had a remarkably gentle temperament. Local fishermen did not sail out to hunt for whales but only caught those that washed ashore or were stranded. This did not happen every day and the local fishermen were shocked to find a whale washed ashore in their bay areas.[72]

Also, fishermen feared the punishment of the gods for catching a whale even if the stranded whale was doomed to die on its own. For these reasons, each time they found a whale washed ashore, they solemnly mourned its death and created a grave for it. Generally speaking, fishermen in Iyo mourned the deaths of whales mostly out of benevolence for whales, more than out of feelings of guilt.[73]

In contrast, fishermen-whalers in Tosa, who had engaged in classical whaling professionally, did not have the luxury of creating a grave for each of the whales they caught. Instead, local whaling guilds commissioned the local temple to make memorial tablets for groups of whales and to regularly erect memorial stupas for groups of whales. They also donated giant hanging bells and other offerings to local shrines and temples. They did so in order to atone for killing whales and out of fear of punishment by evil spirits, as well as out of benevolence for whales and to console their souls.[74]

Notes

1. Yoshihara Tomokichi, "Kujira no haka" (Graves of Whales), in Tanigawa Ken'ichi, ed., *Nihon minzoku bunka shiryō shūsei* (Collection of Documents on Japanese Folk Culture), Vol. 18, Tokyo: San'ichi-shobō, 1997, 441.
2. Ibid.; "Muroto-shi" (Muroto City), http://www.city.muroto.kochi.jp/hopweb/joho/html/, accessed January 30, 2016.
3. Yoshihara, 434–435.
4. Ibid., 435, 441.
5. Ibid.
6. Ibid., 435; Kumano Taiji-ura hogei-shi hensan-iinkai, ed., *Kumano no Taiji: Kujira ni idomu machi* (Taiji, Kumano: Town That Challenges Whales), Tokyo: Heibonsha, 1965, 80–81.
7. Ibid. (both).
8. Kumano Taiji-ura hogei-shi hensan-iinkai, 81.
9. Ibid.
10. Yoshihara, 435.
11. Ibid., 434–436.
12. Ibid., 434.
13. Ibid.
14. Ibid.

15. Ibid.; "Kujira bunka: Kujira o tomuratta kujira-baka · kujira-zuka nado" (Whale Culture: Whale Graves and Monuments to Mourn the Deaths of Whales), http://www.geocities.co.jp/NatureLand-Sky/3011/kujirahaka.html, accessed November 14, 2015.
16. Ibid. (both).
17. "Shikoku hachijū-hakkasho reijō-kai" (Society of Shikoku Eighty-eight Sacred Places), http://www.88shikokuhenro.jp/kochi/26kongochoji/, accessed September 7, 2015.
18. "Dai-26-ban fudasho Ryūtō-zan Kōmyō-in Kongōchō-ji" (No. 26 Temple, Ryūtō-zan Kōmyō-in Kongōchō Temple), http://www.geocities.jp/ftyk14/henro/koutireisyo/26.htm, accessed December 21, 2015.
19. Ibid.
20. "Kujira bunka: Kujira o tomuratta kujira-baka · kujira-zuka nado"; "Hogei hassen-tō shōryō kuyō-hi 2-ki" ("Two Memorial Cenotaph for the Souls of the Eight Thousand Whales Caught"), October 20, 2010, http://www.asahi-net.or.jp/~rn2h-dimr/ohaka2/12kujira/17sikoku/k_hassen.html; "Izui Moriichi" (Izui Moriichi), August 20, 2011, http://hogei.seesaa.net/article/221343007.html
21. Ibid. (both).
22. Ibid. (both).
23. Ibid. (both).
24. Yoshihara, 435.
25. Ibid.
26. Ibid.
27. "Kujira bunka: Kujira o tomuratta kujira-baka · kujira-zuka nado"; "Shikoku-henro no tabi-nikki" (Journal of Shikoku Pilgrimage), November 9, 2013, http://blog.goo.ne.jp/karekusa_2005/e/60f74fdd592b7fe5f4bdc5efef57d8c1
28. Ibid.
29. Ibid.
30. "Whale Watching," http://guides-japan.com/spots/detail/1112/Whale%20watching, accessed, September 8, 2015.
31. "Kujira bunka: Kujira o tomuratta kujira-baka · kujira-zuka nado."
32. Yoshihara, 436–437; "'Tobiagari-kujira no tsuka' and 'Kujira no tsuka'" ("Grave for Jumping Whale" and "Grave of Whale"), September 3, 2009, http://www.asahi-net.or.jp/~rn2h-dimr/ohaka2/12kujira/17sikoku/e_utidomari.html
33. Yoshihara, 436–437; "Kujira bunka: Kujira o tomuratta kujira-baka · kujira-zuka nado."
34. Ibid. (both).
35. Ibid. (both).
36. "Kujira bunka: Kujira o tomuratta kujira-baka · kujira-zuka nado."

37. Yoshihara, 436–437; "Kujira bunka: Kujira o tomuratta kujira-baka · kujira-zuka nado."
38. Ibid.
39. "Kujira bunka: Kujira o tomuratta kujira-baka · kujira-zuka nado."
40. Ibid.
41. Ibid.; "Hotta Minoru Hamachi-yōshoku-sagyōjō" (Hotta Minoru Hamachi Farm Plant), December 28, 2015, http://www.mapion.co.jp/phonebook/M26026/38203/23832000150/
42. "Kujira bunka: Kujira o tomuratta kujira-baka · kujira-zuka nado."
43. Ibid.
44. Ibid.
45. "Hayanote no kujira-zuka" (Whale Grave at Hayanote), http://yoitokona.jp/kujira.html, accessed December 26, 2015; "Akehama-chō no kujira-zuka 3-ki" (Three Whale Graves in Akehama-chō), September 5, 2003, http://www.asahi-net.or.jp/~rn2h-dimr/ohaka2/12kujira/17sikoku/akehama.html
46. Ibid. (both).
47. Yoshihara, 436–437; "Hayanote no kujira-zuka."
48. Yoshihara, 437; "Akehama-chō-Takayama puchi-tanbō" (Exploring Akehama-chō-Takayama), May 4, 2013, http://blog.livedoor.jp/awatennbou/archives/1832867.html
49. Ibid. (both); Boreal Institute for Northern Studies, ed., *Small-Type Coastal Whaling in Japan: Report of an International Workshop*, Edmonton, Alberta: Boreal Institute for Northern Studies, the University of Alberta, 1988, 56.
50. Yoshihara, 437; "Akehama-chō-Takayama puchi-tanbō."
51. "Akehama-chō no kujira-zuka 3-ki"; "Kujira no o-haka" (Whale Graves), May 22, 2011, http://www.seiyo1400.jp/c/blog/2624
52. Ibid.
53. Yoshihara, 436–438; "Akehama-chō no kujira-zuka 3-ki"; "Kujira no o-haka."
54. Yoshihara, 436–438; "Kujira no o-haka."
55. "Akehama-chō no kujira-zuka 3-ki."
56. "Kujira no o-haka"; "Kujira bunka: Kujira o tomuratta kujira-baka · kujira-zuka nado."
57. Yoshihara, 436–438; "Kujira bunka: Kujira o tomuratta kujira-baka · kujira-zuka nado."
58. Ibid. (both).
59. "Kujira bunka: Kujira o tomuratta kujira-baka · kujira-zuka nado."
60. Ibid.
61. Ibid.
62. Ibid.

63. Ibid.
64. Ibid.
65. Ibid.
66. Ibid.
67. Ibid.
68. Ibid.
69. "Sekizōbutsu-gun" (Group of Memorial Stones), October 21, 2009, http://blog.livedoor.jp/awatennbou/archives/1054771.html
70. Ibid.
71. Yoshihara, 438.
72. Hosokawa Takao, "Kujira ni 'kansha to ikei' no nen" (Feelings of "Appreciation and Respect" for Whales), *Mainichi Shimbun*, October 26, 2010.
73. Ibid.
74. Ibid.

CHAPTER 9

Buddhism, Catholicism, Shintoism, and Whaling Culture in the Kyūshū Region

This chapter visits Kyūshū Island, the southwestern major island of the Japanese archipelago. The east coast of the island, encompassing Fukuoka, Ōita, and Miyazaki prefectures, and the west coast, including Saga and Nagasaki prefectures, present dichotomous whaling cultures. Facing narrow straits such as the Bungo Channel, which separates Kyūshū Island from Shikoku Island, local fishermen on the east coast of Kyūshū Island did not engage in classical whaling. However, there are a plethora of whale graves on the east coast because local communities created graves for whales that had drifted in from the Pacific Ocean and became stranded ashore, as in the case of Ehime prefecture on Shikoku Island on the other side of the Bungo Channel. In turn, the west coast of Kyūshū Island, facing the Sea of Japan, was the seat of the Saikai Classical Whaling Base, where distinctive memorial rites for whales developed and many notable whale graves and monuments have been preserved.

Whale Graves in Fukuoka Prefecture

Proceeding from north to south on the east coast of Kyūshū Island, there are two whale graves in Fukuoka prefecture. One is in Kutsuo, Yukuhashi, on the east coast of Fukuoka prefecture, facing the Sea of Buzen. In August 1903, a 54-foot-long whale drifted into the bay area of Kutsuo. The local fishermen sent for a harpooner in Nagasaki and asked him to drive the whale back to the sea. But the whale would not leave the bay, and the harpooner and the local fishermen caught it, and they later created a grave for it. A

M. Itoh, *The Japanese Culture of Mourning Whales*,
https://doi.org/10.1007/978-981-10-6671-9_9

respectable gravestone stands about 6.7 feet tall and its inscription reads "*Kujira-zuka*" (Whale Grave).[1]

Another whale grave is located in Hakozaki, in the capital city Fukuoka, facing Hakata Bay in the northwestern part of the prefecture. In October 1888, a 39-foot-long whale drifted into Hakata Bay and the local fishermen in Amiya caught it. In order to pray for the soul of the whale, the fishermen buried a part of its spinal bone and created a grave for the whale in the bay area near the local streetcar railroad behind the Kyushu Imperial University School of Agriculture. The gravestone is 2.5 feet in height and 0.9 feet in width and has the inscription "*Kujira no shirushi*" (Specimen of Whale). Then, in 1950, when the streetcar railroad was expanded in the bay area, the grave was transferred to the university campus and the bones of the whale were reinterred there. Local people call it "*Kujira no haka*" (Grave of Whale).[2]

Half a century later, in January 1994, the Hakozaki Fisheries Cooperative Union transferred the whale grave to Amiya Tenjin Shrine, where the Ebisu God (Whale God or God of Fisheries) is enshrined, as the eternal place for the soul of the whale to rest in peace.[3]

Whale Graves and Monuments in Ōita Prefecture

Whale Grave in Bungo-Takada

Traveling south from Fukuoka prefecture to Ōita prefecture, there are a number of whale graves there, because local fishing communities created them for the whales that had drifted into the Bungo Channel from the Pacific Ocean and stranded ashore. The gravestones stand on top of the hill overlooking the scenic coastline along the Bungo Channel so that the souls of the whales could "see" the ocean where they had swum. Moving from north to south along the coast, there is a whale grave in Kuresaki, Bungo-Takada on the Kunisaki peninsula in the northeastern part of Ōita prefecture. It was made in May 1843 and stands in the local Ishibe Common Cemetery. The inscription reads:

> "*Namu Amida-butsu*" (Faith in Amida Buddha), "May Tenpō 14" (May 1843 in the Gregorian calendar), contributed by Tamenari Sadakatsu[4]

In addition, in 1888, local fishermen donated a whale *ema* with the painting of a whaling scene to the local Wakamiya Hachiman Shrine in appreciation of a catch of a big whale; however, it was lost when the shrine was transferred after a flood destroyed the building in 1955.[5]

Whale Graves in Ōita

On the southern side of the Saganoseki peninsula in Usuki Bay, there are two whale graves at Isshakunoya, Ōita, the capital city of Ōita prefecture. One was made in March 1855. The tall gravestone is 6.2 feet in height × 2.5 feet in width × 1.3 feet in depth, and the inscription reads, "*Namu Amida-butsu*" (Faith in Amida Buddha). Another whale grave was made in December 1888. The gravestone is 5.6 feet in height × 2.2 feet in depth × 0.08 in depth. The inscription reads:

> "*Hōgō Shaku-Kaisei*" (His afterlife name, follower of Amida Buddha, Kaisei); died on December 7, 1888; the secular name, "*Kujira*"(Whale)[6]

Hōgō refers to an afterlife Buddhist name in the Jōdo-shinshū (True Pure Land School Buddhism), whereas other schools of Buddhism call the posthumous name "*kaimyō.*" In turn, "*shaku*" refers to a title for *hōgō* given to a male person, and denotes a follower of Amida Buddha. The afterlife name of this whale, "*Kaisei*," means everyone can become (a Buddha), while his secular name, "*Kujira*," means whale. This *hōgō* suggests that local fishermen commissioned the local temple head priest to confer the afterlife name on the whale. In fact, *a kakō-chō* (*lit.*, "book of the past"; a death register) has been preserved at the local Kyōson Temple where the head priest recorded the afterlife names of whales.[7]

Whale Graves in Usuki

Going south along Usuki Bay one arrives at Usuki. The city is famous for the scores of *magai-butsu* (statues of Buddha carved on a huge rock or cliff), which resemble "Stone Buddhas" at grotto temples in China and India, such as the Yungang Grottoes in Shanxi province in China. While many stone Buddhas in the region have decayed and been destroyed in the course of time, as many as fifty-nine Usuki Stone Buddhas are designated as Japan's National Treasures. Reflecting the striking Buddhist culture of the locale, whale graves abound in fishing communities in Usuki. One of them stands in Sashiu, Usuki. This grave was made for a 78-foot-long whale that washed ashore near an island off Usuki Bay on January 15, 1886. The fishermen buried a part of the whale's skull and five coins in the grave. The inscription on the gravestone reads:

"*Taigei zengyo kuyō-tō*" (Memorial Stupa for Big Whale Good Fish)[8]

On the back of the gravestone are listed the names of *amimoto* (owners of fishing equipment and boats), fishermen's guilds, and the village, as the monument's financial contributors. "Big Whale Good Fish" is the afterlife Buddhist name of the whale and suggests that the fishermen and the community commissioned the local temple head priest to confer an afterlife name on the whale. The local Ōhashi Temple has actually preserved the death register where the head priest recorded the afterlife names of whales.[9]

Elsewhere in Usuki, there are two whale graves in Ōhama; one in the bay village Haso and another in Matsugahana. The grave in Ōhama-Haso was made for a whale that was stranded ashore while chasing a school of sardines. The gravestone is 3.1 feet in height × 0.9 feet in width. The inscription on the gravestone reads:

"*Shakuni kujira*" (Female follower of Amida Buddha, Whale); May 18, 1894[10]

"*Shakuni*" refers to a title for a *hōgō* given to a female person, while "*shaku*" is given to a male person. Accordingly, the inscription reveals that the grave was made for a female whale. This *hōgō* also suggests that the fishermen and/or the community commissioned the local temple head priest to confer an afterlife name on the whale. In turn, another whale grave in Ōhama-Matsugahana was created for a sperm whale that was caught on February 27, 1944. A *torii* (a sacred entrance gate of a shrine) with a framed wooden board engraved with "*Kujira-zuka*" (Whale Grave) stands on the north side of the grave. The gravestone is 3.4 feet × 0.7 feet in width and its inscription reads: "*Kujira no haka*" (Grave of Whale).[11]

Whale Grave in Ōtomari, Usuki

The most notable memorial rites for whales in Usuki are preserved in Ōtomari, Usuki. Local history records that the Ōtomari folks incurred an enormous debt when they repaired the Ōtomari port in 1868. Then, on the dawn of February 1, 1870, a 105-foot-long male fin whale was stranded ashore. The community sold the whale and used the profit for the debt payment. On February 16, the villagers placed the remains of the whale, including the eyeballs and some of the bones, in a fine pot, buried the pot in a grave on top of the hill looking over the port, and held a grand funeral service for the whale there. In so doing, the community invited as many as fifteen monks from the local Ōhashi Temple, who conducted a funeral service and *Segaki*, a ritual to pray for ending the suffering of the dead who have no living relatives to take care of their afterlives.[12]

The monks also offered the prayer of *Hyakuman-ben* (chanting the sutra one million times). In addition, the community commissioned the head priest at Ōhashi Temple to accord an afterlife name to the whale as well as to make an *ihai* (memorial tablet) for the whale. The posthumous name was "*Tensei taigei zengyo*" (Reincarnated Big Whale, Good Fish). The head priest also recorded the detailed provenance of the whale in the *kako-chō* (the death register) on February 1, 1870.[13]

Moreover, a year later on February 1, 1871, the local folk erected a magnificent memorial stone for the whale on the first anniversary memorial day. The monument is nine feet tall. Its inscription reads:

> "*Taigeigyo hōtō*" (Memorial Treasure Stupa for Big Whale Fish)[14]

Since then, the community has religiously held memorial services and grand *segaki* rituals for the whale at the grave on February 1, every year, by inviting monks from Ōhashi Temple. The community even conducted the 100-year anniversary memorial service on February 1, 1969 (according to the old Japanese counting system). These elaborate and continuous memorial rites given solely for the whale reflect the unique religious mores of Usuki, as well as the depth of appreciation and benevolence of the local community toward the whale that saved it from financial difficulty.[15]

There is also a stone statue called "*Kujira Jizō*" (Guardian Deity of Motherhood and Unborn Children for Whales) in Itachiya, Usuki, west of Ōtomari. The statue has an engraving, "*Okina*" (Elder) and local people state that the statue was made for the person who invented a spear to catch swordfish; however, an accurate provenance is unknown.[16]

Whale Shrines in Ōita Prefecture

In addition, there is a whale shrine, Ebisu Shrine, in another bay village, Nakatsu-ura, Usuki. On September 10, 1882, local fishermen caught a 48-foot-long whale that was stranded ashore chasing after a school of sardines, and sold the whale. In appreciation and atonement, local *amimoto* (owners of fishing equipment and boats) made a grave for the whale and erected a gravestone at Ebisu Shrine. Its inscription reads "Kujira-jinja" (Whale Shrine), with the date and the names of *amimoto* as financial contributors for the monument, including Hiramatsu Yashirō and Higashi Tokujirō. Remarkably, there are as many as 89 Ebisu Shrines that worship the Ebisu God in Ōita prefecture alone, out of which 55 have the same name, Ebisu-jinja (Ebisu Shrine). This also suggests the religious mores of this region.[17]

Whale Graves in Kamiura, Saiki

Going down south to Saiki (not Saeki), two whale graves are located in Azamui, Kamiura-machi (which was incorporated into Saiki in March 2005) on Saiki Bay. One was made in February 1888 and the other in December 1907. The older gravestone is 4.4 feet height × 0.8 feet in width × 0.8 feet in depth and has the inscription with the date and name of the fishing guild:

> "*Namu Amida-butsu Geigyo-tō*" (Faith in Amida Buddha Memorial Stupa for Whale Fish).[18]

In turn, the whale grave made in 1907 was for a 42- or 66-foot-long whale that a villager found stranded ashore at dawn. The gravestone is 4.7 feet height × 0.8 feet in width × 0.8 feet in depth and the inscription reads:

> "*Namu Amida-butsu, Geigyo-baka*" (Faith in Amida Buddha, Grave for Whale Fish)[19]

The two whale graves stand side by side. Since then, whenever the villagers found a whale on the shore, they said to each other, "The whale came to pay a visit to the whale graves." This saying derives from the Japanese tradition of visiting the graves of their ancestors and family members on religions days, including the Vernal Equinox, Bon Week (the week of August 15), and the Autumnal Equinox, as well as on their individual memorial days. The villagers felt that whales swam to the shore and visited the whale graves in order to pay tribute to the souls of their family.[20]

Whale Grave in Miyazaki Prefecture

Moving south on the east coast to Miyazaki prefecture, there is a whale grave at Tobashira Shrine at Tobashira Park in Aburatsu, in Nichinan. The gravestone is 4.7 feet in height and 3.3 feet in width. The inscription reads:

> Erecting this memorial stone in order to console the soul of a whale[21]

Local history has it that a few hundred years ago Aburatsu was a small island, being separated from communities on the coast. When a typhoon hit the island, the villagers of Aburatsu did not have any food to eat and were on the verge of starvation. Then, a large whale ran aground on the eastern shore of Aburatsu and the fishermen caught the whale. The villagers consumed the whale as a blessing of the gods, but they felt deeply sorry when they found a fetus inside in the body of the whale. They carefully buried the fetus as well as the eyeballs of the mother whale at a local Shōgyō Temple and also erected a gravestone. The words "to console the soul of a whale" on the engraving suggests that the local villagers feared punishments from the gods for catching the pregnant whale and sought atonement and forgiveness. The gravestone was transferred to a different site when the temple was rebuilt and currently stands next to a memorial cenotaph for those who were lost at sea, at Tobashira Park in Aburatsu.[22]

Meanwhile, the villagers of Aburatsu did not forget that the whale saved the lives of their ancestors a few hundred years earlier and expressed their appreciation by making *kujira mochi* ("whale cake"; an eight-inch long rice cake in the shape of a whale with red-bean filling) during the Bon Festival every year. (As per Buddhist tradition, the Japanese believe that the souls of the ancestors return to Earth during Bon Week and celebrate this holy week by conducting memorial services and offering foods to the souls of the ancestors). The local community of Aburatsu has retained the tradition of making "whale cakes" and exchanges them among the community members on Bon Week, to this day.[23]

* * *

History of Saikai Classical Coastal Whaling Base

Next, it is time to move to the west coast of Kyūshū Island. With such fishing communities in Yobuko and Ogawa Island (in current Saga prefecture) and Ikitsuki Island (in current Hirado, Nagasaki prefecture), the Saikai

Classical Whaling Base ("saikai" literally means "west sea") on northwestern Kyūshū constituted one of the four major classical whaling bases in Japan, along with the Kishū Classical Whaling Base in Taiji, the Chōshū Kitaura Classical Whaling Base in Nagato, and the Tosa Classical Whaling Base in Muroto. Accordingly, there are many whale graves and monuments in the region. Today, the Saikai Classical Coastal Whaling Base roughly corresponds with the Saikai National Park of Japan, with scenic islands jeweled with historic buildings, including numerous Catholic churches.[24]

Nakao Jirōzaemon (1583–April 1663), a native of Nakao, Ōmura (in current Nagasaki prefecture), began classical whaling in Kyūshū in earnest. He traveled far to Taiji, Kishū province, to learn the classical manual harpoon method. Upon returning home, he organized a whaling guild in Ōmura, with the authorization of the fourth lord of Hizen-Ōmura province, Ōmura Suminaga (September 1636–September 1706), whose great-great-grandfather was the first Christian-converted lord in Japan, Ōmura Sumitada (1533–June 1587). Meanwhile, Nakao Jirōzaemon expanded his business into Iki, Ikitsuki, and Gotō Islands, and was given the surname "Fukazawa" by Ōmura Suminaga. Thus, Nakao became Fukazawa Gidayū I, Katsukiyo.[25]

Fukazawa Gidayū not only made a considerable financial contribution to the Hizen-Ōmura province, but also founded a new temple, Tendai School Buddhism Shōrin-zan, Daijō-in Enyū Temple, in current Kusaba-Kōji, Ōmura, in order to atone for his sins and pray for the souls of whales. In 1684, his son, Fukazawa Gidayū II, Katsuyuki, also went to Taiji and learned the net method from Wada Kakuemon Yoriharu in person, and began his operations in Arikawa on Gotō Islands. This is the beginning of classical net whaling in Kyūshū. Fukazawa Gidayū II was buried in Uonome on the Gotō Islands, instead of his family temple, Hakuryū-zan Chōan Temple in Ōmura, because he died in Uonome.[26]

Whale Graves and Monuments in Saga Prefecture

Classical Whaling in Yobuko

The earliest recorded history of classical whaling in Yobuko, in current Karatsu, Saga prefecture, dates back to 1594. In 1716, local whaling guild heads, the Nakagawa clan and the Nakao clan, introduced the net method and established their operations in Yobuko and Ogawa-jima, an island off north Yobuko. However, as the Nakao clan expanded its operations as far as

Chikuzen-Himejima in the east and the Gotō Islands in the west, the Nakagawa clan ended its operations in the 1760s and the Nakao family became one of the dominant classical whalers in Kyūshū. The Nakao clan prospered for eight generations until 1878 when the Ogawa-jima Whaling Guild Company (which became the Ogawa-jima Whaling Company in 1888) took over the whaling guild of Nakao Zenpei in the eighth generation of the Nakao clan.[27]

Founding of Zen Buddhism Ryūshō-in Temple in Yobuko

Meanwhile, the third generation in the Nakao clan, Nakao Jinroku III, Shigehiro (1731/1732–August 1781), subscribed to Buddhism more deeply than his father and grandfather. One day in 1755, he had a dream in which a pregnant whale appeared and begged him to save her life. The next day, he found that his men had caught a whale. This whale turned out to be carrying a fetus. Nakao mourned the death of the whale to the extent that he spent all the profits from the whale on restoring a local temple, Sōtō School Zen Buddhism Ryūshō-in Temple, in Yobuko in 1755. Since then, the Nakao family has regularly conducted funeral services and memorial services for whales at the temple.[28]

The famous chronicle of classical whaling in the locale, *Ogawa-jima geigei gassen* ("Whale Battles at Ogawa Island"), compiled by Shūhōtei Riyū in 1840, records that crewmen on their whale boats repeatedly chanted *nenbutsu* (the Buddhist prayer) of *Namu Amida-butsu* (Faith in Amida Buddha) at the death of each whale after the catch. The fishermen especially felt deep sorrow and guilt when they caught a mother whale and her calf.[29]

The chronicle also describes the grand funeral service for whales held at Ryūshō-in Temple, to which the Nakao family invited a number of monks from other local temples. The Nagao family commissioned the head priest of Ryūshō-in to accord a *hōgō* (afterlife name for True Pure Land School Buddhism) to each whale and to record each name on a memorial wooden tablet with the date of death of the whale. At the funeral service, while monks were reading each whale's afterlife name aloud, members of the whaling guild sat solemnly and prayed for the souls of the whales.[30]

After finishing the chanting prayer, monks, along with guild members, went to the shore and sent the wooden memorial tablets off to sea, along with other offerings, so that the souls of the whales would reach the Pure Land and attain the enlightenment of Amida Buddha. This memorial rite is based on the local tradition called *shōrō nagashi* (*lit.*, "sending off the spirits to the sea"), in which people send off a boat carrying the souls of their ancestors to sea so that the souls can return to the Pure Land of Paradise (the world of the dead) on the Bon memorial day in August. The local folk in Nagasaki and Saga prefectures have preserved this tradition and still perform this ritual each year on August 15.[31]

The Chronicle of "Whale Battles at Ogawa Island" has illustrations that depict scenes of memorial rites for whales conducted by Ryūshō-in Temple. One of them is entitled *Ryūshō zenji ni oite kujira kuyō no zu* ("Picture of Memorial Service for Whales at Ryūshō Zen Temple"). This picture shows the head monk of Ryūshō-in Temple performing a memorial service for whales caught during the season, joined by his fellow monks from Sainen Temple, Gankai Temple, Chōtoku Temple, Kan'non Temple, and Ryūsen Temple. While the six monks chanted prayers aloud (the figures in Photograph 9.1 appear to be laughing, but they are actually chanting prayers), representatives of the whaling guild, including a hazashi in formal kimono attire, and guests, including shrine head priests (*bettō*), chanted prayers together (see Photograph 9.1).[32]

Another interesting illustration compiled in the Chronicle of "Whale Battles at Ogawa Island" is *Kujira kuyō nagashi kanjō no zu* ("Picture of Memorial Rite of Transferring the Souls of Whales by Sending Them off to the Sea"). This picture depicts a scene of a monk sending off boats carrying wooden memorial tablets for whales and offerings to sea, as whaling guild heads watched them (see Photograph 9.2).[33]

Folklore Legend of Whale Spirit in Yobuko

There is also another folklore legend of the spirit of a mother whale in Yobuko. One day, a *hazashi* ("blade thruster"; a leader of whale boats, see Chap. 2) of the Nakao Whaling Guild had a dream in which a spirit of a whale with a calf appeared and begged him not to catch them because they were on their way to make a pilgrimage to Benten Island, where Benten or the Benzaiten Goddess (see Chap. 2) was enshrined. Next day, the *hazashi* found a mother whale with a calf in the sea. Remembering the dream, he hesitated to catch them, but his fellow fishermen urged him to do so. In the evening, when he returned home, a harpoon hung at the entrance of his house fell on his child, killing the child instantly. Shocked by this accident, he lost his mind and waded into the sea to die, carrying the dead child. This kind of folklore legend has also been told in other classical whaling communities in Japan as a cautionary tale.[34]

Photograph 9.1 "Picture of Memorial Service for Whales at Ryūshō Zen Temple" in Chronicle of "Whale Battles at Ogawa Island," Yobuko, Saga prefecture, 1840

Photograph 9.2 "Picture of Memorial Rite of Transferring the Souls of Whales by Sending Them off to the Sea" in Chronicle of "Whale Battles at Ogawa Island," Yobuko, Saga prefecture, 1840

Whale Grave in Yobuko

Regarding monuments for whales in Yobuko, a whale grave exists at a common cemetery in Kimizuka, among many graves for the local residents, including those of the Nakao family. The gravestone for this whale grave is about 2.6 feet in height, but stands four feet tall with its pedestal. The inscriptions read:

> "*Namu Amida-butsu*" (Faith in Amida Buddha) [on the front side]
> "*Banrei geigei jōbutsu*" (For all the Souls of Whales to become a Buddha in heaven) [on the back side]
> Nakagawa Yoshibee Shigetsugu (the local whaling guild head) [on the side]
> "July 15, Enkyō 5" (August 8, 1748 in the Gregorian calendar; the Enkyō era officially ended on July 12, and was succeeded by the Kan'en era) [on the side][35]

Memorial Stupas for Whales at Ryūshō-in Temple

In 1813, another whaling guild head, Matsuo Karoku, created a magnificent monument for whales at Ryūshō-in Temple in Yobuko. His great grandfather, Matsuo Kōdayū, who was a founding head of the local whaling guild, had previously erected a memorial stupa for whales in 1714. However, this old stupa had decayed and was damaged with time, and Matsuo Karoku decided to create a new stupa for whales in commemoration of the one-hundred-year anniversary of the creation of the original one, as well as the occasion of his eighty-eighth birthday.[36]

The new memorial stupa is made in the style of a *gorin-tō* (a five-tiered stupa), which consists of five tiers of stones curved in different shapes. They are from bottom to top, a stone in the shape of a cube (representing earth), a sphere (representing water), a pyramid (representing fire), a hemisphere (representing wind), and a round "treasure" (representing sky). It was a popular style for stupas for Shingon and other schools of Buddhism in Japan.[37]

The most unique feature of this five-tiered stupa at Ryūshō-in Temple is demonstrated on the third tier from the bottom. The third tier is a stone in the shape of a cylinder and reliefs of the *Six Jizō-bosatsu* (Guardian Deities of Motherhood and Unborn Children) are engraved on the surface. Each of the *Six Jizō-bosatsu* represents one of the six treasures that are believed to save the living and the dead from the Six Paths and Three Realms of Desires and Karmic Rebirth. The reliefs of the *Six Jizō-bosatsu* suggest that Matsuo Karoku wanted to atone for catching pregnant whales carrying fetuses, as well as whales in general. A stone in the shape of a cube on the first tier at the bottom has many inscriptions. The inscription on the front reads:

> "*Geigei kuyō*" (Memorial for Whales)[38]

Inscriptions on the other sides record the detailed provenance of the old and new memorial stupas.[39]

Memorial Stupa for One Thousand Whales at Ryūshō-in Temple

In August 1831, the seventh generation of the Nakao family, Nakao Jinroku VII, Yūnosuke, created another memorial stupa at the family temple, Ryūshō-in Temple, as the number of whales that the Nakao Whaling Guild had caught since its founding reached one thousand. The monument is in a shape of a cube, and has inscriptions:

> "*Geigei senbon kuyō-tō*" (Memorial Stupa for One Thousand Whales) [on the front]
> "Nakao VII, Yūnosuke Guild" [on the right side]
> "July Tenpō 2" (August 8–September 5, 1831 in the Gregorian calendar) [on the left side][40]

Also, on the pedestal, thirty-one names of guild members are inscribed, suggesting that each of them had made a sizable financial contribution to the temple to create this monument, in order to atone for the sins their guild had accumulated since its foundation.[41]

Memorial Stupa for Whales on Ogawa Island

Classical whaling on Ogawa-jima, an island north of Yobuko, began in 1594. The Nakagawa family and the Nakao family moved to Ogawa Island and established a solid whaling base on the island in 1751. The whaling on the island was chronicled in *Ogawa-jima geigei gassen* (Whale Battles at Ogawa-jima), written in 1840. The Nakao clan created a monument for whales on the island in 1863. It was erected in the local Kan'non-dō, a temple that enshrines *Kan'non-bosatsu* (the Goddess of Mercy, one of the guardian deities of Amida Buddha). The inscription reads:

> "*Geigei kuyō-tō*" (Memorial Stupa for Whales), "March Bunkyū 3" (April 18–May 17, 1863)[42]

This monument is made in the style of *funa-gata* (boat-shape) and has a relief of *Jizō-bosatsu* engraved on the center of the boat. This suggests that it was dedicated to the souls of pregnant whales, as well as other whales. In addition, there is a whale grave and memorial stone in the graveyard of the Nakao clan in the temple compounds, but their provenances are not well documented.[43]

Whale Temple in Karatsu

There is also a respectable memorial stone for whales at Pure Land School Buddhism Jōshō-zan Kōju-in Temple in Tōbō, Karatsu, near Yobuko. The temple is located at the Tōbō fishing port in Karatsu Bay. The local folklore legend has it that this old monument was created for the souls of whales. However, other than the fact that it was created in the early Edo period (1603–1868), its correct provenance is unknown, because the inscription is difficult to read and the meaning is unclear. What is known is that this venerable monument has quietly looked over the Tōbō fishing port and Karatsu Bay for four centuries. Next to it stands a newer stone statue of *Jizō-bosatsu*, wearing a red bib. The two stone monuments look as if they are watching over fishermen and whales at sea.[44]

Whale Shrines in Yobuko

Local fishermen also worshiped the Ebisu God (Whale God and God of Fisheries) and made sizable financial contributions and supported local shrines in Yobuko. Among them, local whalers patronized Yobuko Hachiman Shrine (which has a stone statue of the Ebisu God), Atago Shrine, Tenman Shrine, and San Shrine. For instance, the Yobuko Whaling Guild donated a large *waniguchi* (a flattened hanging bell with a slit at the bottom that makes it look like the mouth of a shark, see Chap. 7) to Hachiman Shrine and made other offerings. Also, the local whaling guild donated a stone *torii* (sacred entrance gate) to Tajima Shrine on Ogawa Island.[45]

In turn, the Ogawa Island Whaling Company donated huge stone lanterns to Atago Shrine, with unique statues of a pair of *komainu* (guardian dogs) standing on their hind legs on top of each lantern at the entrance of the shrine (the guardian dogs at other shrines are usually sitting, not standing). Atago Shrine commands the whole view of Yobuko fishing port, and the guardian dogs appear to stand as tall as possible, as if they were watching over fishermen in the ocean far away. They might symbolize the hearts of the wives of the fishermen-whalers, who saw their husbands off at the shrine and prayed for their safe sailing.[46]

The worship of the Ebisu God had deeply penetrated into the local fishing communities, and statues of Ebisu God not only abound on the shores of Yobuko, but each household also has a miniature shrine with a statue of the Ebisu God. The community has also preserved a whale processing song. Today, Yobuko has transformed itself from a classical whaling base into a squid fishing town, and "Yobuko squid" is famous nationwide.[47]

Statue of Ebisu God Riding on Whale on Kashiwa Island

There is also an interesting stone statue of the Ebisu God on Kashiwa Island (Kashiwa literally means "god-gathering") off the northeastern tip of the Higashi-Matsuura peninsula (administratively it belongs to Minato-chō, Karatsu). The fishermen on the island used to engage in classical whaling, employing the net method, as well as fishing Japanese yellowtails (*buri*), and worshiped the Ebisu God. The local whaling guild commissioned a statue of the Ebisu God in a unique style, with him riding on a whale and holding a yellowtail in his arm. The inscription has the date of creation, February 1895, and the name of the whaling guild. By worshiping the Stone Statue of the Ebisu God Riding on Whale, the local whalers prayed for safety at sea and a good catch.[48]

Whale Shrine in Nagoya, Karatsu

Moving to the northwestern tip of the Higashi-Matsuura peninsula, there is another stone statue of the Ebisu God at Furusato Shrine in Nagoya, Karatsu. Nagoya used to be a stronghold of the Matsura clan (spelled different from Matsuura, as in the Higashi-Matsuura peninsula) and there stood Nagoya Castle (currently, only the stone foundation remains, but the former site of the castle is designated as a Special Historic Site of Japan). The stone statue of the Ebisu God in Nagoya is about 3.0 feet in height and 1.5 feet in width, and is accompanied by a smaller statue of *Jizō-bosatsu*. The inscription on the statue reads:

"Hazashi, Ushitarō" and "September 20, Ansei 3" (October 18, 1856)[49]

Commenting on this *hazashi* ("blade thruster") by the name of Ushitarō, a blog writer states, "Being such a leading professional of classical whaling, he must have felt a strong need for atonement and prayer for the souls of whales and donated the statue of the Ebisu God to the shrine, as a memorial for whales."[50]

Further, going south to Imari (the seat of prized Imari Porcelain), two monuments for dolphins stand on Taniga Island on the west side of Imari Bay (in current Haze, Yamashiro-chō). One was erected in December 1921 and another in February 1943 in commemoration of a big catch. Both of the monuments have a stone statue of the Ebisu God on top and the inscription, "*Iruka dai-myōjin*" (Great God of Dolphins), at the bottom. These monuments suggest that the local community suffered from a food shortage or financial difficulties and caught dolphins. They feared punishment from the gods and erected the monuments in dedication to the god of dolphins and asked for forgiveness.[51]

Whale Graves and Monuments in Nagasaki Prefecture

Whale Grave in Isahaya

Leaving Saga prefecture and moving west to Nagasaki prefecture, there is a large whale grave in Nakaoki-chō, Isahaya, the third largest city in Nagasaki prefecture, after Nagasaki and Sasebo. This fishing village did not engage in classical coastal whaling because it was located at the foot of Isahaya Bay that faces the Sea of Ariake, a large inland sea in western Kyūshū. The year 1884 was a lean one for the local fishermen in Isahaya, and they could not make proper preparations for the New Year celebration (the most important and longest holidays in traditional Japan). Then, a large whale drifted into Isahaya Bay in January 1885. The fishermen caught it and sold it, which solved their financial crisis. In appreciation and atonement, the local fishermen created a grave for the whale. Its large stone monument is covered with moss and looks venerable. The larger inscription on the upper side of the stone and the smaller one on the lower side read:

"*Kujira-tō*" ("Memorial Stupa for a Whale")
"Contributed by Isahaya Bay Fishermen" and "January Meiji 18" (January 1885)[52]

However, more than a hundred years later, most of the local people seem to have forgotten about this whale grave and few people visit the grave. Other stone monuments stand next to it, including a stone statue of the Ebisu God and a memorial stone for fish made by the Isayama Fisheries Cooperative Union.[53]

Whale Graves on Iki Island

Moving to the north to Iki Island in the Sea of Japan, there are two whale graves in Iki. Two are located at Ebisu-ura (*lit.*, "Ebisu bay"), Ashibe-chō, on the eastern coast of the island. One whale grave has a respectable eight-foot-tall stone stupa made of granite. It is in the style of a "*kasa-tōba*" (*lit.*, "hat stupa" because it is decorated with a roof that looks like a hat) and stands solemnly on top of the hill, looking over the bay to the east. This monument was created in April 1717 by a local whaling guild in order to pray for the souls of mother whales and their calves and fetuses that they had caught. The inscriptions read:

"*Hō-dokushō daijō-myōten issenbu*" (Dedicating the chanting of one thousand copies of the Sutra of Mahayana Buddhism) [on the front]
"May whales quickly attain the enlightenment of Buddha" [on the right side]
"May this guild continue to be blessed," [on the left side]
"March Kyōho 2" (April 12–May 10, 1717) [on the back]
[Five names of guild members are inscribed on its pedestal][54]

There is a smaller and older grave next to this one. The inscription on the smaller gravestone has faded away and is illegible, but local people state that it was made for a mother whale and her calf or fetus. Ebisu-ura used to be a disassembling site in classical whaling. The fishermen were surprised and saddened when they found a fetus in the whale they had caught. The same flowers were offered in each of the pair of vases placed in front of the two graves on October 11, 2009. The two whale graves in Ebisu-ura suggest the depth of remorse and atonement of the local whalers for catching mother whales and pregnant whales, as well as the benevolence of the local residents today. In fact, local people have conducted memorial services for the whales annually. The community also retained a whale song.[55]

New Whale Monument on Iki Island

In recent years, a new monument for whales was created in Katsumoto-chō on the northern coast of Iki Island. The Nagatori clan, which had served the last classical whaling guild head in Katsumoto-ura, erected a memorial tower for whales in March 2000, in order to pray for their souls, as well as those of the crewmen who had worked for the family. The original name of the Nagatori clan was Harada. As whales stopped passing off the bay of Katsumoto in the mid-nineteenth century, the tenth lord of Hirado province, Matsura Hiromu (May 1791–July 1867) who had received sizable financial contributions from Harada, gave Harada a new surname "Nagatori" (*lit.*, "to catch forever") in June 1864.[56]

Nevertheless, as modern Western whalers began to deplete the Japan Grounds (see Chap. 2) at the time, the Nagatori clan ended whaling operations. This stone monument is about 4.3 feet in height (including its pedestal) and 3.5 feet in width. The inscription reads: "*Kujira kuyō-tō*" (Memorial Stupa for Whales). The memorial stone is flanked by supplementary stones with inscriptions of the provenance of this monument and of the family.[57]

In addition, there used to be a large bronze hanging bell, with a relief of a whale engraved on it, at Kongō-zan Kaihō Temple in Numatsu village (current Gōno-ura) on the southwestern coast of Iki Island. This suggests that a local whaling guild made a sizable financial contribution to the temple for commissioning the hanging bell. Local history records that pirates from Tsushima Island off the Sea of Japan (located in the Korea/Tsushima Strait that divides Japan and Korea) sailed to Iki Island one day and stole a gilded-bronze statue of Buddha and the whale hanging bell from Kaihō Temple. The pirates could not sail back with the stolen items as they had planned, so they sunk the statue of Buddha and the whale hanging bell in the sea and sailed away in a hurry. Later, the statue of Buddha was caught in a fishing net and was restored, but the whale hanging bell was never found.[58]

Moreover, there is a memorial stone for dolphins on Tatsunoshima, an island north of Katsumoto-chō. An approximately five-foot-long stone monument, with the inscription "*Iruka irei no hi*" (Memorial Cenotaph to Pray for the Souls of Dolphins), overlooks the ocean. This suggest that the fishermen on the island suffered from a poor catch of fish had caught dolphins and erected the monument in atonement.[59]

Hirado Island

Hirado Island is located at the northwestern tip of Kyūshū Island with its main city Hirado, the westernmost city in the Kyūshū Region. (Hirado Island was connected to Kyūshū via a land route in 1977 with the completion of the Hirado Grand Bridge.) Historically, Hirado prospered during the sixteenth and early seventeenth century as a port of call for foreign ships—such as those of Chinese, Dutch, English, and Portuguese origin—until 1639 when the Tokugawa shogunate government enforced the seclusionist *sakoku* policy and closed Japanese ports, with the exception of Nagasaki, in order to shut down Catholic and other foreign influences.[60]

Subsequently in 1641, the Tokugawa government transferred the Dutch Trade House from Hirado to Nagasaki, in order to control and monopolize foreign trade, as well as to weaken the power of local lords in Kyūshū. As a result, the revenue of Hirado province drastically decreased. In order to counter the Tokugawa shogunate government's deliberate measures to weaken Hirado province, the fourth lord of the province, Matsura Shigenobu (April 1622–November 1703), developed various alternative sources of revenue, including classical whaling.[61]

Meanwhile, a number of local lords in the northwestern region of Kyūshū, as well as farmers and fishermen, had converted to Catholicism due to influence of Portuguese and Spanish missionaries. In turn, the Tokugawa shogunate government banned Catholicism in fear of its influence and persecuted the Japanese who had converted to Catholicism. Many Japanese converts refused to reconvert to Buddhism and became martyrs, while others became "Hidden Christians." The latter secretly practiced Christianity in caves on Hirado Island, Ikitsuki Island, and the Gotō Islands, like Christians who had practiced in the catacombs in ancient Rome.[62]

The Japanese novelist Endō Shūsaku wrote a compelling story of the Hidden Christians in *Chinmoku* (Silence) in 1966. In April 2016, the grave of an Italian missionary, who was one of the models for the protagonists of *Chinmoku*, was discovered in the former site of the *Kirishitan-yashiki* (the prison for Christians). A DNA test confirmed this. In December 2016, a Hollywood film based on the story and directed by Martin Scorsese was released in the United States. Today, among the many churches in the region, one Catholic church on Hirado Island and three churches on the

Gotō Islands were designated as Cultural Properties of Japan by the Japanese government, while sacred places and villages on Hirado Island are designated collectively as an Important Cultural Landscape. Along with the Ōura Cathedral and related facilities in Nagasaki, which are designated as a National Treasure of Japan, these churches and historic sites constitute part of the "Churches and Christian Sites in Nagasaki" of UNESCO World Heritage, with this status in a tentative list as of 2015.[63]

Whale Graves in Iwanoue-chō, Hirado

There are two whale graves with memorial stupas at the Shingon Buddhism Kōya-san Saikyō Temple in Iwanoue-chō in the northern part of Hirado Island. The first lord of Hirado province, Matsura Shigenobu (1549–July 1614, his great grandson, the forth lord, assumed the same name), was a devout follower of Shingon School Buddhism and forcibly took over the local Sōtō School Zen Buddhism Shōon-in Temple in 1607. Earlier, this temple had been a Shingon School Buddhism temple. The local legend has it that the founder of Saint Kūkai (Kōbō-daishi, 774–April 834) visited Hirado in 806 on his way back from his training as a Buddhist monk in Tang-dynasty China, and upon returning to Japan conducted the first ritual of making offerings into a consecrated fire at this temple.[64]

In recourse to this old story, Matsura Shigenobu tried to reconvert this Zen Buddhism temple into a Shingon Buddhism temple (or transfer the Zen temple somewhere else), but the temple head priest Ryūdon refused. As a punishment, Matsura burned the temple. Ryūdon and his disciple Eitetsu died in the fire, protecting the principal sacred statue of the temple, the *Yakushi-nyorai* (Buddha of Healing and Medicine). Afterward, Matsura founded Saikyō Temple there and emplaced a statue of Shingon Buddhism founder Saint Kūkai, which was custom made in Kyoto. Saikyō Temple is referred to as "Kōya-san of the West" (Saint Kūkai founded the original Kōya-san Kongōbu Temple in current Wakayama prefecture), and houses many treasures, including a "picture of the nirvana of Buddha" (*nehan-zu*) that is designated as an Important Cultural Property of Japan.[65]

The two whale graves were created in the front yard of the innermost sacred house (*okuno-in*) built behind the main house (*honden*) of Saikyō Temple. The older whale grave was made in August 1695. The memorial stone for this grave is in the shape of a cube, with a round-shaped roof (like the roof of a mosque). It is about 3.8 feet in height and 1.3 feet in width. The inscriptions read:

"*Geigei kuyō-tō*" (Memorial Stupa for Whales) [on the front]
"Early July Genroku 8" (August 1695) [on the side]
Tominaga Lin'emon (a local whaling guild head) [on the back][66]

In April 1938, the local whaling union, the Uematsu-gumi, created a new whale grave next to the old one in the same graveyard. The new gravestone is in the style of a *gorin-tō* (a five-tiered stupa) and is about 5.3 feet in height, including its rock foundation, and 1.6 feet in width. The inscriptions read:

"*Geirei kuyō-tō*" (Memorial stupa for the souls of whales)
Yoshizumi Jinpei (the whaling union head) on the back[67]

Five-Tiered Stone Stupa for Whales in Mae-Tsuyoshi, Hirado Island

In addition, there is a magnificent five-tiered stone stupa for whales at Chōsen Temple in Mae-Tsuyoshi in the southern part on Hirado Island. The Oda Whaling Guild that was based in Ojika Island (a northern island of the Gotō Islands) established a whaling base in Mae-Tsuyoshi in 1692 and operated it until 1859. In April Genbun 4 (May 8–June 5, 1739), the eighth head priest of Chōsen Temple, Kaijun, collected donations in order to create a monument to pray for the souls of whales. This five-tiered stupa is made of a total of 14 stones and stands 18 feet tall. The first tier (at the bottom) has an open space in the middle, and inside this space statues of the Abundant Treasures Buddha and the Amida Buddha (both eight inches tall in a sitting posture) are housed.[68]

Interestingly, moreover, the names of various Buddhas of *Hokke-kyō* (Lotus Sutra Buddhism) are inscribed on the surface of the stone. A huge ginkgo tree stands next to the monument, as if to guard the stupa. This monument represents stone buildings made in Hirado during the mid-Edo period, and was designated as a Folk Cultural Property of Nagasaki Prefecture in September 1984.[69]

Whale Grave on Azuchi-Ōshima Island

In recent years, a whale grave was discovered on Azuchi-Ōshima, a small island north of Hirado Island (the island was incorporated into Hirado city in October 2005). Local history records that a man by the name of

Yoshirobee from Kishū (where Taiji is located) began classical whaling on the island in the 1620s. Also, Kawamoto Magoshichiemon Yoshitoshi from Nagahama, Ōmi (current Shiga prefecture, east of Kyoto), settled down on the island and was appointed as its first political officer by the first Hirado-province lord Matsura Shigenobu. Kawamoto's son was given the surname "Imoto" by the lord and thus changed his name to Imoto.[70]

As noted earlier, Hirado province lost enormous revenue when the Tokugawa shogunate government closed the Hirado port and transferred the Dutch Trade Office to Nagasaki in 1641. Consequently, in 1661 (or 1664) the fourth Hirado-province lord Matsura Shigenobu ordered the third Azuchi-Ōshima political officer, Imoto Yashichizaemon Yoshinobu (the grandson of Kawamoto Magoshichiemon Yoshitoshi), to engage in classical whaling. (The Hirado city official website states that Imoto began classical whaling in 1661, while Yoshihara states that Imoto began in 1664.)[71]

The whaling operations succeeded beyond Imoto's expectations, and he made an enormous financial contribution to Hirado province. He also restored True Pure Land School Shinkyō Temple in 1667, which was originally founded in 1615. The whale grave stands among the graves for the Imoto clan in the compounds of this temple. The gravestone is about 4.1 feet in height × 1.7 feet in width. Its inscription reads:

> Thirty-three Reincarnations of Whales, July 1692, Kawakubo Heiemon[72]

The Imoto clan engaged in classical whaling until 1726 when the fifth generation closed down the business (the whaling equipment had been burned in a fire in 1725). Imoto Kiyoshi, a descendant of the Imoto clan and a local inn owner on the island, found the whale grave in his family graveyard at Shinkyō Temple. The village of Kōno-ura on Azuchi-Ōshima has retained the historic buildings and streets of the whaling community for a hundred years, and in June 2008 the area was designated as one of the Important Preservation Districts for Groups of Traditional Buildings of Japan.[73]

Ikitsuki Island

Surprisingly, Ikitsuki Island, a small island of less than 6.4 square miles, northwest of Hirado Island, was one of the major stations for the Saikai Classical Whaling Base. The island was incorporated into Hirado city in

October 2005. (Ikitsuki Island was connected to Hirado Island via a land route in 1991 with the completion of the Ikitsuki Grand Bridge, which is the longest continuous truss bridge in the world.) In the late seventeenth century, Tatamiya Matazaemon Kagemasa, a descendant of Yamagata Saburobee, who was a retainer of Takeda Shingen, the famous warrior of the State of Kai (current Yamanashi prefecture), settled down on the island.[74]

His son, Tatamiya Matazaemon II, Masakatsu, organized a whaling guild, with his father-in-law, Tanaka Chōdayū, and began classical whaling in 1725. The guild was successful and Masakatsu expanded his operations into a large area encompassing Mae-Tsuyoshi on Hirado Island and Tsushima Island in the Tsushima Strait (the boundary with Korea), and as far as Kayoi-ura in Chōshū (current Yamaguchi prefecture). Masakatsu became a financial backbone of Hirado province, so that the province lord gave him the family name Masutomi upon his death. Thus, he became Masutomi Matazaemon I, Masakatsu, posthumously.[75]

The fourth generation, Masutomi Matazaemon IV, Masazane, even conducted whale research at the northern frontier, *Ezo-chi* (current Hokkaidō), by order of the Tokugawa shogunate government. The Masutomi clan reached its zenith during the fifth generation of Masutomi Matazaemon V, Masahiro, when the Masutomi clan managed five whaling guilds, commanding more than 200 whale boats and 3000 crewmen. It was the largest whaling operation in Japan during the Edo period. However, the number of whales migrating off the Sea of Japan decreased in the mid-1840s and the family business began to decline during the time of Masutomi Matazaemon VI, Masataka.[76]

In the 1850s, American whalers began to deplete the Japan Grounds and whales stopped coming. The Japanese fishermen were shocked to find bodies of whales killed by foreign harpoons and those of skinned whales washed ashore. Consequently, Masutomi Matazaemon VII, Masatada, ended whaling in 1873 (or 1874 depending on the source), after 150 years of operations by his clan. In total, the Masutomi clan caught 21,790 whales in about 150 years between 1725 and 1873. In comparison, the American whalers caught 53,877 whales in the Japanese grounds in about 100 years, from the early nineteenth century to the early twentieth century. On this, Yoshihara Tomokichi writes that it is quite surprising that the Masutomi clan caught that many whales using vastly inferior and inefficient methods compared to those of the Americans.[77]

Statue of Ebisu God with Whale on Ikitsuki Island

There is no whale grave preserved on Ikitsuki Island, probably because most of the islanders were Catholic (Hidden Christians) and, instead of Buddhist temples, Catholic churches dominated the small island. With this religious background, the Masutomi Whaling Guild founded Misaki Shrine near its whale storage building at Misaki Bay at the northern end of the island, and also created a statue of *Kujira Ebisu* (the Ebisu God with Whale) at the facility. The statue is made of tuff and is about 1.2 feet in height. It is housed in a small shrine, which is currently managed by Hakusan Shrine, because no one manages Misaki Shrine today.[78]

The main shrine of Misaki Shrine does not exist, and only the *torii* gate and smaller shrines housing statues of various gods still exist.) This statue of the Ebisu God is significant in its unique style, in which the Ebisu God, holding a sea bream in his arm, is sitting on a whale caught in a fishing net. Because of this unusual style, this statue is considered a valuable monument in the study of the relationship between the Ebisu God and classical Japanese whalers.[79]

Ikitsuki Township Museum

Significantly, this island has a fine folk museum, "Ikitsuki-chō hakubutsukan · Shima no yakata" (Ikitsuki Township Museum · A Mansion on an Island), which houses artifacts and equipment of classical whaling, as well as a skeleton of a North Pacific right whale. One of the treasures of the museum is a two-volume *Isana-tori e-kotoba* ("Illustrations of Whaling"), consisting of 41 woodblock print illustrations that depict various scenes of classical whaling, whale boats, and whales in striking detail. It was commissioned by Masutomi Matazaemon V, Masahiro in 1829 and was completed in 1832 (an addendum was published later).[80]

"Illustrations of Whaling" recorded that the affection of mother whales toward their calves was deep, and humpbacks had an especially intense and heartbreaking bond. This awareness and feelings of guilt for killing mother whales and their calves produced several legends about whales on Gotō Islands (examined below).[81]

The Ikitsuki Township Museum also exhibits many mementoes of the Hidden Christians on the island, such as the robe of a priest and a rosary, as well as the replica of a house of hidden worship (many Hidden Christians threw themselves from the cliffs of the island, in order to avoid execution, as did those on Hirado Island, the Gotō Islands, and elsewhere in the region). In spite of (and because of) this history of persecution, most of the islanders have remained Catholic and have retained the unique form of rites that the Hidden Christians devised in order to disguise themselves as Buddhists. In recent years, the township created a large statue of *Dai-gyoran-kan'on* (the Goddess of Mercy for Fish), which was completed in April 1980. At 60 feet tall (in a sitting posture) and weighing 150 metric tons, it is the largest bronze statue in Japan. (The statue sits outside like the Grand Buddha of Kamakura.)[82]

Gotō Islands

Moving southwest, one reaches a large island group, the Gotō (*lit.*, "five islands") Islands. Being part of the "Churches and Christian Sites in Nagasaki," the islands are adorned with Catholic churches, three of which are designated as Important Cultural Properties of Japan. Also, being part of the Saikai Classical Whaling Base, there are many whale graves and monuments on the islands. Two of the monuments for whales are designated as Historic Sites of a local town, Shin-Kami-Gotō-chō.[83]

Whale Graves on Uku Island

On the Gotō Islands, there are two whale graves on the northernmost island, Uku Island (the island was incorporated into Sasebo, Nagasaki prefecture, in March 2006). One whale grave is located at Tōkō Temple in the Taira district on Uku Island. Tōkō Temple was the official temple of the local lord of the Uku clan. Local legend has it that Taira no Iemori (1123 circa–April 1149, a younger brother of the ruler in the late Heian period, Taira no Kiyomori) settled down on Uku Island and became Uku Jirō Iemori (the founder of the Uku clan). The inscription on the gravestone reads:

"*Geikon kuyō-tō*" (Memorial Stupa for the Souls of Whales)[84]

There is another whale grave in the cemetery. It is made of a natural stone and is covered with moss. It has the inscription "Grave for One Thousand Whales, December 1924." To this day, Tōkō Temple has conducted a memorial service for the souls of whales every spring.[85]

In addition, there is Ebisu Shrine in Horikawa in the Taira district, where a lookout post for classical coastal whaling used to stand. Enshrining the Kujira Ebisu (Whale God/God of Fisheries), the local fishermen have conducted the Koshiki Festival at the shrine, in order to pray for a good catch of fish, in November every year to this day.[86]

Folklore Legend of Yamada Monkurō on Uku Island

There is a famous folklore legend about the whaling guild head, Yamada Monkurō, and a whale, on Uku Island. He was in the third generation from Yamada Shigebee, who had moved to Uku Island from Satsuma province (current Kagoshima prefecture in southern Kyūshū) and had founded a whaling guild on Uku Island. 1715 was a lean year for the Yamada Whaling Guild. The crewmen suffered financially and hoped to catch a whale before the new year (the Japanese tried to clear debts at the end of each year so that they could celebrate a new year debt-free), but no whale swam into their fishing territory. Then on the night of January 21, 1716 (in the lunar calendar), Yamada Monkurō had a strange dream, in which a blue whale with a calf said, "We are on our way to the pilgrimage to Daihō Temple. Please do not catch us until we finish the pilgrimage."[87]

Shingon Buddhism Daihō Temple (founded in 701) was located on Tamanoura, Fukue Island (the southernmost island of the Gotō Islands). A legend has it that the Shingon Buddhism founder, Saint Kūkai (Kōb-ō-daishi), was hit by a typhoon in 806 on his way back from Tang-dynasty China and took refuge on Fukue Island and taught Shingon Buddhism at Daihō Temple. Because of this history, Daihō Temple is referred to as the "Kōya-san of the West." (Saikyō Temple on Hirado Island is also referred to as the "Kōya-san of the West.")[88]

True to what the whale said in Yamada Monkurō's dream, January 22 was a religious holiday for Saint Kūkai. Yamada therefore told his crewmen about his dream and told them not to catch a blue whale with a calf. Nevertheless, despite Yamada's order, when a whale watcher spotted a 200-foot-long blue whale with a calf, the crewmen sailed off and tried to catch it. The mother whale fought back. Then, a storm with freezing sleet hit the crewmen and they were shipwrecked. A total of seventy-two crewmen went missing or washed ashore. After this tragedy, Yamada quit whaling and began a sake brewery business. There is a different version of this folklore legend on Fukue Island (examined below).[89]

As if to prove this legend, there is a memorial stupa to pray for the souls of the seventy-two shipwrecked crewmen on Uku Island. The inscriptions on the monument read:

> "*Namu Amida-butsu*. For the souls of the seventy-two crewmen to attain the enlightenment of Amida Buddha..." [on the front]
> "January 22, Shōtoku 6" (February 18, 1716) [on the side]
> "Hazashi, Yamada Monkurō Guild" (the name of the contributor) [on the side][90]

However, scholars have noted discrepancies in this legend. For one thing, Yamada Monkurō actually died in 1714 (two years before the tragic accident), and there is a grave for Yamada Monkurō on Uku Island. Also, no graves for the seventy-two shipwrecked crewmen made in Shōtoku 6 (1716) have been found.[91]

Whale Grave on Ojika Island

Traveling directly south from Uku Island, there is a whale grave on Ojika Island (the island belongs to Ojika-chō, Kita-Mastuura county, administratively). The Oda clan from Iki Island and the Asai clan from Ōmura (near Nagasaki) organized their whaling guilds on Ojika Island and conducted classical coastal whaling there. In the second generation of the Oda family, Oda Denbee created a whale grave in the family's graveyard in Obuchi, Fuefuki, on the island in 1695. The inscription on the memorial stone reads:

> "*Tame geigei bodai*" (For Whales to Attain the Enlightenment of Buddha)[92]

This monument was designated a Tangible Cultural Property (Building) of Ojika town in March 1982.[93]

There is a local folklore legend about this whale grave. One night Oda Denbee had a dream, in which a huge 200-foot-long whale appeared and said, "I am passing by your bay tomorrow morning on my way to a pilgrimage to Ise Grand Shrine. Please do not catch me. If you let me accomplish my pilgrimage, I will have myself caught by you on my way back from the pilgrimage." Next morning Oda tried to tell his crewmen about his dream, but it was too late. The whaleboats had already sailed off to the sea. A huge whale was caught that day. Oda mourned the death of the whale and created the grave and erected a monument for the whale.[94]

There is another version of this folklore legend. In this variation, a fine lady appeared in the dream of Oda Denbee and told him "a 200-foot-long blue whale will pass through your fishing grounds at dawn tomorrow on his way to the Ryūgū Palace (the undersea palace of the Dragon God of the Sea). Since the whale is delivering a message to the palace, please do not catch him. You can catch him on his way back if you must." As Oda was going to say, "I understand," the lady had already disappeared from his dream.[95]

The next morning Oda Denbee sent a messenger to the fishing grounds to tell his crewmen not to catch a blue whale, but it was too late. They had already caught a 200-foot-long blue whale. Oda regretted that he could not keep his promise in the dream and felt sorry for the whale. Being afraid of punishment from the gods of the sea, Oda decided to quit whaling and began a business of seaweed harvesting and sake brewing. He also cultivated Hario Island and created a rice field there. His new business prospered.[96]

Whale Grave in Nama, Aokata, Nakadōri Island

Moving south to Nakadōri Island, the second largest island of the Gotō Islands (the island was incorporated into a newly created township, Shin-Kami-Gotō-chō, Minami-Matsuura county, in August 2004). Nakadōri Island has as many as twenty-nine Catholic churches—one of them is designated as an Important Cultural Property of Japan—and many of the residents on the island are Catholics. There is a whale grave in Nama, Aokata. Aokata port is a natural haven for fishing boats and is also famous for *Aokata-monjo* (Aokata Archives), which chronicle the history of the Aokata clan that had joined the influential Matura-tō, an alliance of maritime forces (*suigun*) ruled by the Matsura clan.[97]

The whale grave stands in a corner of the graveyard that the locals call the Cemetery for Military Officers. Local history records that classical whalers created the whale grave in order to atone for the sin of killing whales. They did so by commissioning local monks to perform the prayer of *Ichiji-isseki-kyō* (*lit.*, "Sutra of one-letter-for-one-stone"). This refers to a form of memorial rite in Lotus Sutra Buddhism in Japan, in which monks copied down the sutra—one Chinese character per flat pebble—and buried them around the grave.[98]

The "Sutra of one-letter-for-one-stone" for whales in Aokata was originally buried in Fukuju-an Temple. A few hundred years later, in 1900, when several gravestones in Fukuju-an fell down, locals found there many pebbles with a Chinese character written on each of them. They were the "Sutra of one-letter-for-one-stone" for whales made during the Edo period. The local people refer to this grave as "*Kuyō-baka*" (Memorial Grave) for whales. They collected the pebbles and reburied them next to the grave of a past lord of the region, the Aokata clan.[99]

Whale Grave in Arikawa, Nakadōri Island

Arikawa, on the eastern tip of Nakadōri Island, was a center for classical whaling on the Gotō Islands. The first whaling guild in Arikawa was organized in 1598 by order of the first lord of Hizen-Fukue province (also called Gotō province), Gotō Harumasa (1548–April 1612, of the twenty-first generation of the Gotō clan; his Christian name was Louis). In October 1712, the Arikawa village head, Eguchi Jin'emon Masatoshi, who had developed fisheries in Arikawa Bay, made a whale grave on top of Mt. Kujira-mi (*lit.*, "whale watching"; a location of a sentry post), looking over the ocean. (The current Arikawa town hall stands on the site of the former residence of Eguchi Jin'emon Masatoshi.) The inscription of the gravestone reads:

> Memorial for the 1,312 whales this bay village caught from 1691 to 1712, erected in October 1712 by Eguchi Jin'emon Masatoshi[100]

The Arikawa community also erected a new monument with the inscription "*Kujira kuyō-tō*" (Memorial Stupa for Whales), with a detailed history of classical whaling in Arikawa.[101]

Whale Shrine in Arikawa, Nakadōri Island

There is also a whale shrine in Arikawa on Nakadōri Island. Kaidō Shrine at the foot of Mt. Kujira-mi is dedicated to the God of the Sea and the Ebisu God (the Whale God), and has a gate made of a pair of whale jawbones. Local history records that the local communities had a series of accidents involving the drowning of adults and children from 1617 to 1619. In order to prevent such accidents in the sea in the future, a local official built this shrine in 1619, to pray to the *Kaidō-shin* (the God of Children at the Sea). Hence this shrine is called Kaidō Shrine. At the shrine, there is a memorial stone for a family that had engaged in "dolphin fisheries," suggesting that the family had saved local people on the verge of starvation.[102]

Whale Graves in Uonome, Nakadōri Island

There are two whale graves in Maruo in Uonome, on the northern tip of Nakadōri Island. They stand side by side on top of a hill called Mt. Kuyō (*lit.*, "memorial"). (Uonome village became Shin-Uonome-chō in September 1956, and then was incorporated into a newly created town, Shin-Kami-Gotō-chō in August 2004.) The whale grave on the left is called the "Grave for One Hundred and Six Whales," which was made by the Uonome-ura Whaling Guild in 1696. The gravestone is about 6.3 feet in height and 1.2 feet in width, and has the inscriptions:

> "*Namu Amida-butsu*" (Faith in Amida Buddha), along with other words of a prayer in the Buddhist sutra in smaller letters [on the front]
> "Respectfully dedicating a stone stupa in order to pray that the souls of whales may rest in peace, as well as respectfully dedicating the prayer of one-character-for-one-stone of Volume One of the three-volume Pure Land Sutras" [on the side]
> "June 15, Genroku 9" (July 13, 1696), the Uonome Guild [on the side][103]

The guild created another whale grave in April 1774. Its gravestone is about 2.2 feet in height and 16 feet in width, and curved in the style of a *chikara-ishi* (*lit.*, "strength stone"; a heavy rock curved in a rounded shape used in temples and shrines in Japan). The inscription reads:

> "*Kujira hyappon kuyō*" (Memorial for One Hundred Whales)
> "March Kanpō 4" (April 1744)[104]

To the right of these two whale graves stands a grave of the founder of the Uonome-ura Whaling Guild, Fukazawa Gidayū II, Katsuyuki, who died on July 16, Genroku 7 (September 5, 1694). As mentioned earlier, his father, Nakao Jirōzaemon, learned the classical harpooning method in Taiji, expanded his business to Iki Island, Ikitsuki Island, and the Gotō Islands, and was given the surname "Fukazawa." Then, Fukazawa Gidayū II Katsuyuki learned the net method in Taiji, began his operations in Katsumoto on Iki Island, and then organized his whaling guild in Uonome. The grave of Fukazawa Gidayū II, Katsuyuki and the two whale graves are housed in a wooden structure like a small shrine, flanked by a statue of the Ebisu God and a statue of *Jizō-bosatsu*. To this day, three local temples have taken turns in conducting annual memorial services for whales and for Fukazawa on January 5.[105]

Folklore Legend of Yogorō and Whale

There is a folklore legend of a whaling guild head Yogorō and a pregnant whale in Uonome. A whale appeared in the dream of Yogorō and begged him not to catch her, because she was on her way to give birth in the ocean. However, fishermen sailed out and tried to catch the whale. Then, suddenly a storm hit the boat and sank it. Realizing that it was the punishment of the gods of the sea, Yogorō erected a monument for the whale and kept praying for the soul of the whale until his death.[106]

Whale Graves on Fukue Island

Moving further south to Fukue Island, the largest island of the Gotō Islands (the island was incorporated into a newly created city, Gotō, in August 2004), there are fourteen Catholic churches, as well as two whale graves: one in Kojima, Tomie and another in Kurose. The gravestone in Kojima, Tomie, looks very old, but the large inscription is still legible; it reads, "*Kuyō-baka*" (Memorial Grave). The local people say that it is a whale grave, but an accurate provenance is unknown. Also, a beautiful memorial stone curved in the shape of an oval stands next to the gravestone. It is about 6.6 feet in height and 4.3 feet in width.[107]

In turn, another whale grave in Kurose has the following inscription on its gravestone:

> "*Tame geigei jōbutsu*" (For the Sake of Whales to Become Buddhas); "April Genroku 3" (May 9–June 6, 1690); "Fuji Shin'emon, Yuasa-ura, Kishū"[108]

This inscription indicates that the person who made the financial contribution to create the grave, Fuji Shin'emon, was a native of Yuasa Bay, Kishū (current Wakayama prefecture, where Taiji is located). The local history of the Gotō Islands also records that a native of Yuasa, by the name of Shōsuke, organized a whaling guild in Arikawa on Nakadōri Island in 1626. These records suggest a strong connection between the Kishū Classical Whaling Base and the Saikai Classical Whaling Base, just as Fukazawa Gidayū I, Katsukiyo and Fukazawa Gidayū II, Katsuyuki each went to Taiji to learn the classical harpoon method and net method, respectively, from the Wada family of Taiji in person.[109]

Folklore Legend of Yamada Monkurō on Fukue Island

The folklore legend of Yamada Monkurō and a blue whale on Fukue Island is slightly different from that on Uku Island introduced earlier. According to the Fukue-Island version, despite Yamada's order, his crewmen and his son, Jinkurō, sailed off to catch a 200-foot-long blue whale on its way to the pilgrimage to Daihō Temple on Fukue Island. The whale fought back and attacked the whaleboats, killing all the crewmen. Watching his son being killed, Yamada Monkurō lost his mind, challenged the whale, and was also killed. The whale then died and washed ashore. The villagers referred to the whale as the "Monkurō Whale," and since then have performed a whale dance in dedication to the Monkurō Whale on January 16 (in the lunar calendar) every year.[110]

The existence of different versions of the folklore legend of Yamada Monkurō and the whale, as well as similar folklore legends on the Gotō Islands, suggests that local fishermen-whalers took these stories as a cautionary tale for not catching blue whales, which were simply too large and thus beyond the capacity of Japanese classical whalers equipped only with rudimentary tools.

Notes

1. "Kujira bunka: Kujira o tomuratta kujira-baka · kujira-zuka nado" (Whale Culture: Whale Graves and Monuments to Mourn the Deaths of Whales), http://www.geocities.co.jp/NatureLand-Sky/3011/kujirahaka.html, accessed November 14, 2015.
2. "Kujira no shirushi" (Specimen of Whale), http://www.asahi-net.or.jp/~rn2h-dimr/ohaka2/12kujira/18kyusyu/hakata_sirusi.html, accessed November 18, 2015.
3. Ibid.
4. "Kujira bunka: Kujira o tomuratta kujira-baka · kujira-zuka nado"; Yoshihara Tomokichi, "Kujira no haka" (Graves of Whales), in Tanigawa Ken'ichi, ed., *Nihon minzoku bunka shiryō shūsei* (Collection of Documents on Japanese Folk Culture), Vol. 18, Tokyo: San'ichi-shobō, 1997, 441.
5. Ibid.
6. "Kujira bunka: Kujira o tomuratta kujira-baka · kujira-zuka nado."

7. Ibid.
8. Ibid.; Yoshihara, 438–440.
9. Ibid. (both).
10. Yoshihara, 439–440; "Kujira bunka: Kujira o tomuratta kujira-baka · kujira-zuka nado."
11. Ibid. (both).
12. Ibid. (both).
13. Ibid. (both).
14. Ibid. (both).
15. Ibid. (both).
16. "Kujira bunka: Kujira o tomuratta kujira-baka · kujira-zuka nado."
17. Ibid.; Yoshihara, 440; "Zenkoku Ebisu-ōkami hōsai-sha ichiran" (List of Shrines in Japan that Worship Ebisu Grand God), http://nishinomiyaebisu.web.fc2.com/todouhuken/menu/table/43table.html, accessed January 18, 2016.
18. Yoshihara, 438, 440; "Kujira bunka: Kujira o tomuratta kujira-baka · kujira-zuka nado."
19. Ibid. (both).
20. Ibid. (both).
21. Yoshihara,441–443; "Kujira bunka: Kujira o tomuratta kujira-baka · kujira-zuka nado."
22. Ibid. (both).
23. Yoshihara, 442.
24. "Saikai National Park," https://www.env.go.jp/en/nature/nps/park/saikai/index.html, accessed November 23, 2015.
25. Yoshihara, 452 and 470–471; Kumano Taiji-ura hogei-shi hensan-iinkai, ed., *Kumano no Taiji: Kujira ni idomu machi* (Taiji, Kumano: Town That Challenged Whales), Tokyo: Heibonsha, 1965, 81–82. For detail of Christians in Kyūshū see, John Breen and Mark Williams, eds., *Japan and Christianity: Impacts and Responses*, New York: St. Martin's Press, 1996.
26. Ibid.
27. Yoshihara, 452–453; "Hogei-ōkoku 'Yobuko'" (Whaling Kingdom: "Yobuko"), http://www.geocities.co.jp/NatureLand-Sky/3011/yobuko.html, accessed November 28, 2015.
28. Yoshihara, 452–454.
29. Ibid.; "Nihon no kujira bunka: Kujira to shinkō" (Japanese Whale Culture: Whales and Beliefs), http://www.geocities.co.jp/NatureLand-Sky/3011/kujira-to-sinkou.html, accessed November 23, 2015; Kon'no Tsutomu, *Kaneko Misuzu futatabi* (Revisiting Kaneko Misuzu), Tokyo: Shōgakukan, 2007, pbk., 2011, 214–215. The word "*geigei*" refers to male and female whales. It is pronounced "*keigei*" in Kayoi, Yamaguchi prefecture, see Chap. 7.

30. Yoshihara, 452–454; “Nihon no kujira bunka: Kujira to shinkō.”
31. Ibid. (both).
32. Ibid. (both).
33. Ibid. (both).
34. Yoshihara, 473; “Nihon no kujira bunka: Kujira ni kakawaru densetsu · itsuwa” (Japanese Whale Culture: Folklore Legends and Anecdotes Regarding Whales), http://www.geocities.co.jp/NatureLand-Sky/3011/33kujira-densetu.html, accessed December 23, 2015.
35. Yoshihara, 451–452.
36. Ibid., 451, 453–454.
37. Ibid.
38. Ibid.; Makikawa Takanosuke, “Yobuko Ryūshō-in to Kujira kuyō-tō” (Yobuko Ryūshō-in Temple and Memorial Stupa for Whales), *Niji* (publication of Karatsu City), No. 24, 1965, 13–17.
39. Ibid. (both).
40. Yoshihara, 450–451.
41. Ibid.
42. Ibid., 453–454; “Hogei-ōkoku ‘Yobuko.’”
43. Ibid. (both).
44. Nakazono Shigeo, “Kujira kuyō-hi” (Memorial Stupas for Whales), unpublished and unnumbered photo collection provided to author, 2012; “Kōju-in” (Kōju-in), April 12, 2015, http://hakataboy.com/temple/Saga/Karatsu/KoujyuIN/m.html
45. “Hogei-ōkoku ‘Yobuko.’”
46. Ibid.
47. Ibid.; Boreal Institute for Northern Studies, ed., *Small-Type Coastal Whaling in Japan: Report of an International Workshop*, Edmonton, Alberta: Boreal Institute for Northern Studies, the University of Alberta, 1988, 56.
48. Nakazono; “Kashiwa-jima repōto” (Kashiwa Island Report), December 24, 2004, http://www.scs.kyushu-u.ac.jp/~hatt/report04/saga/kasiwa.htm
49. “Furusato-jinja no sekizō ebisu-zō” (Stone Statue of Ebisu God at Furusato Shrine), September 2, 2014, http://www.asahi-net.or.jp/~rn2h-dimr/ohaka2/12kujira/18kyusyu/sk_furusato1.htm
50. Ibid.
51. Yoshihara, 454.
52. “Kujira bunka: Kujira o tomuratta kujira-baka · kujira-zuka nado.”
53. Ibid.
54. Ibid.; Yoshihara, 454.
55. Ibid. (both); Boreal Institute for Northern Studies, 56.
56. “Kujira bunka: Kujira o tomuratta kujira-baka · kujira-zuka nado”; “Nagatori-ke no kujira kuyō-tō” (Memorial Tower for Whales of Nagatori

Clan), September 1, 2014, http://www.asahi-net.or.jp/~rn2h-dimr/ohaka2/12kujira/18kyusyu/ni_nagatorike.htm
57. Ibid.
58. Yoshihara, 456.
59. "Kujira bunka: Kujira o tomuratta kujira-baka · kujira-zuka nado"; "Tatsunoshima jōriku tanken" (Landing and Exploring Tatsunoshima Island), http://www.yado.co.jp/kankou/nagasaki/iki/tatunosima/tatunosima2.htm, accessed November 20, 2015.
60. "Hirado-shi" (Hirado City), http://www.city.hirado.nagasaki.jp, accessed November 21, 2015.
61. Ibid.
62. "Churches and Christian Sites in Nagasaki," http://whc.unesco.org/en/tentativelists/5096/, accessed November 9, 2015.
63. Ibid.; "'Kirishitan-yashiki-ato' kara jinkotsu" (Bones Found at former Site of the Prison for Christians), *Asahi Shimbun*, April 4, 2016, http://www.asahi.com/articles/ASJ444WNRJ44UTIL024.html?iref=comtop_rnavi_arank_nr02
64. Yoshihara, 458; "Saikyō-ji" (Saikyō Temple), May 10, 2011, http://kyushu-tabi-gourmet.seesaa.net/article/200193253.html
65. Ibid. (both).
66. Yoshihara, 457–458; "Geigei kuyō-tō to Geirei kuyō-tō" (Memorial Tower for Whales and Memorial Tower for the Souls of Whales), September 4, 2014, http://www.asahi-net.or.jp/~rn2h-dimr/ohaka2/12kujira/18kyusyu/nh_saikyouji.htm
67. Ibid. (both).
68. "Chōsen-ji no Kujira kuyō sekizō gojū-no-tō" (Five-tiered Memorial Stone Stupa for Whales at Chōsen Temple), July 20, 2008, http://blogs.yahoo.co.jp/misakimichi/41823718.html; "Nagasaki-ken no bunkazai" (Cultural Properties of Nagasaki Prefecture), http://www.pref.nagasaki.jp/bunkaDB/index.html, accessed November 9, 2015.
69. Ibid. (both).
70. Yoshihara, 456–457; "Kōno-ura no rekishi" (History of Kōno-ura), http://www.city.hirado.nagasaki.jp/city/info/prev.asp?fol_id=14152, accessed November 5, 2015.
71. Ibid. (both).
72. Yoshihara, 456–457; "Kōno-ura no machi-nami" (Streets of Kōno-ura), June 9, 2008, http://www.city.hirado.nagasaki.jp/city/info/prev.asp?fol_id=14084
73. Ibid. (both).
74. Yoshihara, 458–459; "Ikituki-chō hakubutsukan · Shima no yakata" (Ikitsuki Township Museum · House on Island), http://www.hira-shin.jp/shimanoyakata/, accessed February 12, 2016.

75. Ibid. (both).
76. Yoshihara, 459, 465; "Ikituki-chō hakubutsukan · Shima no yakata."
77. Yoshihara, 458–459; "Ikituki-chō hakubutsukan · Shima no yakata."
78. "Kujira bunka: Kujira o tomuratta kujira-baka · kujira-zuka nado"; "Misaki-ura hogei-nayaba" (Whale Storage at Bay of Misaki), http://www.yado.co.jp/kankou/nagasaki/ikitukis/hogeinayaba/index.htm, accessed February 15, 2016.
79. Ibid.
80. "Ikituki-chō hakubutsukan · Shima no yakata"; Nakazono Shigeo and Yasunaga Hiroshi, *Kujira-tori e-monogatari* (Picture Tales of Whaling), Fukuoka: Gen-shobō, 2009, 78–129; "*Isana-tori e-kotoba*" (Illustrations of Whaling), September 8, 2014, http://www.photo-make.jp/hm_2/kuzira.html; Kon'no, 214–215.
81. Ibid. (both).
82. "Ikituki-chō hakubutsukan · Shima no yakata."
83. "Shin-Kami-Gotō-chō no bunkazai ichiran" (List of Cultural Properties of Shin-Kami-Gotō-chō), http://k101ow01.town.shinkamigoto.nagasaki.jp/faqhp/20/bunkazai.pdf, accessed November 10, 2015.
84. Yamakawa Tōru, "Kujira ronsō o hajimeru-mae ni: '*Kujira wa darenomonoka*'" (Before Debating Whaling: "Who Owns Whales?"), March 26, 2009, http://business.nikkeibp.co.jp/article/life/20090323/189756/
85. Ibid.
86. "Ukujima" (Uku Island), http://www.ukujima.com/tourist/taira.html#taira05, accessed November 18, 2015.
87. Yoshihara, 473–475; "Nagasaki 'Saikai no kujira densetsu · minwa'" ("Legends and Folklore Tales about Whales" in Nagasaki), http://www.geocities.co.jp/NatureLand-Sky/3011/33kujira-densetu4.html, accessed November 14, 2015.
88. Ibid. (both).
89. Ibid. (both).
90. Yoshihara, 420, 468, 474.
91. Ibid.
92. "Kujira bunka: Kujira o tomuratta kujira-baka · kujira-zuka nado"; "Ojika-chō bunka-zai shōkai" (Introduction to Cultural Properties of Ojika Town), http://www.ojika.net/index2.html, accessed November 16, 2015.
93. Ibid.
94. "Nagasaki 'Saikai no kujira densetsu · minwa.'"
95. Ibid.
96. Ibid.

97. Yoshihara, 468–469; "*Aokata-monjo*" (Aokata Archives), March 1, 2013, http://www.city.nagasaki.lg.jp/shimin/190001/192001/p000591.html
98. Ibid. (both).
99. Ibid. (both).
100. Yoshihara, 469.
101. Ibid.
102. Ibid., 432, 469.
103. Ibid., 470.
104. Ibid.
105. Ibid., 452 and 470–471.
106. "Nihon no kujira bunka: Kujira ni kakawaru densetsu · itsuwa."
107. Yoshihara, 471; Nakazono.
108. Yoshihara, 471–472; "Kujira bunka: Kujira o tomuratta kujira-baka · kujira-zuka nado."
109. Ibid. (both).
110. "Nagasaki 'Saikai no kujira densetsu · minwa.'"

CHAPTER 10

Conclusion

The Japanese culture of mourning the death of whales is a reflection of Japanese religious beliefs. Historically, they believed in indigenous animism which stressed that everything—living and nonliving—has a soul. They found deities in everything and worshiped them. Consequently, the notion of reverence for life in general, as well as the worship of various gods in ancient Shinto mythology (like the many gods in Greek mythology), penetrated the daily life of the Japanese. To this, the belief in Buddhism that proscribed killing living things was added, and the notion of benevolence for both life and death, as well as atonement for killing living things for one's own living, permeated the Japanese tradition. Thus, as a result of the syncretism of Shintoism and Buddhism, the Japanese prayed to the gods in worship and wish-making, and simultaneously prayed for the souls of the dead in mourning and atonement (see Chap. 2).

With this cultural and religious background, Japanese fishermen and whalers engaged in classical coastal whaling. They were deeply religious because their lives were at the mercy of the vagaries of the weather and they needed a strong vehicle of prayer. In the absence of modern scientific knowledge, fishermen-whalers were superstitious and feared punishments of the gods for catching whales that were considered gods of the sea, or messengers of the gods, in many regions. Therefore, they worshiped the Ebisu God and the Hiruko God and sought for their blessings and protection, as well as for forgiveness. They also subscribed to Buddhism and tried to atone for their deeds of killing whales. Kawamura Shigeyoshi, the head

M. Itoh, *The Japanese Culture of Mourning Whales*,
https://doi.org/10.1007/978-981-10-6671-9_10

priest of Kan'non Temple in Ayukawa, Miyagi prefecture, states that the essence of Buddhism boils down to atonement for killing living things.[1]

Kawamura states that,

> realizing the sad fate of human beings that they have to kill living things in order to sustain their own lives, human beings seek for atonement. In so doing, fishermen-whalers commissioned Buddhist temples to conduct a special funeral service for whales called *segaki*, which was normally given to a person who did not have any living relatives to take care of his/her afterlife. The fishermen-whalers prayed that the souls of whales could go to heaven and become Buddhas, based on the notion of reincarnation that believes a beast or a demon could be reborn as a human being or become a Buddha upon death.[2]

This culture of mourning the death of whales took place in the larger framework of the traditions of mourning the death of fish by Japanese fishermen. Japanese fishermen-whalers felt a stronger sorrow for catching whales because they were the largest and most magnificent creatures in the ocean. Therefore, fishermen-whalers offered various forms of memorial rites for whales comparable to those for human beings. Fishermen-whalers in Kayoi, Yamaguchi prefecture, commissioned Kōgan Temple to accord each whale they caught a Buddhist afterlife name and meticulously recorded it in the death register of parishioners, treating whales as if they were individual human beings. Fishermen-whalers in Taiji, Wakayama prefecture, painted their whale boats with designs patterned after the Picture of the Pure Land of Paradise (the equivalent of heaven), praying that whales could go to the Pure Land of Paradise, and chanted *Namu Amida-butsu* in front of the whale at each catch (see Chaps. 3, 4, 5, 6, 7 and 8; Tables A.1 and A.2).

In turn, fishermen-whalers in Yobuko, Saga prefecture, conducted grand funeral services, including the special memorial rites of Sending off the Souls of Whales to the Sea, as depicted in the pictures at Ryūshō-in Temple. All of these locales have conducted memorial services for whales annually to this day. They were not "pretend prayers" merely conducted to justify the deeds of killing whales (see Chaps. 3, 4, 5, 6, 7, 8 and 9; Table A.2).

Empathy for Whales

Many observers have pointed out the empathy that classical Japanese fishermen-whalers held for whales. For instance, the prominent writer Tsunoda Fusako, who visited Ōmi Island in Nagato, Yamaguchi prefecture, writes that

> the mentality of mourning the deaths of whales symbolized the way the Japanese fishermen-whalers dealt with nature. They did not consider a whale an object of hatred or an object of conflict that lasted until one of the two killed the other, as exemplified by Captain Ahab and Moby Dick. The Japanese fishermen-whalers considered both human beings and whales as being subject to the harsh world of nature. They thought that both human beings and whales lived in the world of *mujō*, a notion in Buddhism that nothing in this world is constant, based on the undeniable fact that death is part of life. They grasped that killing whales was a sad fate for both the fishermen and the whales, and extended their empathy to the whales that had given their lives to human beings.[3]

Similarly, Ehime University Department of Agriculture professor Hosokawa Takao states that

> the Japanese fishermen-whalers comprehended whales as an integral part of nature in which they coexisted. They held whales in awe for their might and respected them, and mourned the deaths of whales. Even non-whaling communities created graves for stray whales that had been stranded ashore and mourned their deaths. In this context, fishermen-whalers considered whales as blessings of the sea and consumed them with great appreciation. They did not waste anything and utilized every usable part of the whales. They did not consider whales as 'commodities.' Instead, they treated whales with veneration and took care of their deaths and their 'afterlives' as if they were human beings. In essence, the appreciation, respect, and sympathy for whales that had sustained the livelihoods of fishermen's villages are the essence of Japanese whaling culture.[4]

The fact that even the villagers in non-whaling communities mourned the deaths of whales and created graves for them suggests genuine empathy for whales, because they did not really have to seek for atonement for killing them. They caught only whales that had accidentally been stranded ashore. Yet, they still extended benevolence for whales, because it concerned their core belief system. The classical whalers shared the same core belief system. After all, most of the whaling guild members were fishermen who grew up in rural communities that observed the same culture and beliefs as those in non-whaling communities. In mourning the deaths of whales, they created collective graves and memorial stupas. The Memorial Stupa for Thirty-Three Whales in Nigishima, Mie prefecture, and the Grave for One Hundred and

Six Whales on the Gotō Islands, Nagasaki Prefecture, are such examples (see Chaps. 3, 4, 5, 6, 7, 8 and 9; Table A.1).

These monuments are proof of the long history of the Japanese culture of mourning the deaths of whales, just as the continued observance of memorial rituals attests to the contemporary vitality of Japanese whale-related ceremonies. In this regard, whale festivals that have been conducted in dedication and appreciation to the Ebisu God and other gods of the sea serve to retain the solidarity among community members in preserving their whaling culture and practices (see Tables A.3 and A.4).[5]

Folklore Legends

In addition, an international workshop on Japanese whaling points out that the ubiquity of similar folklore legends in whaling communities might suggest the wisdom of classical Japanese whalers. The essence and lessons of these stories are basically the same: A spirit of a female whale appears in a dream of a Buddhist monk or a whaling guild head, telling him that she and her calf would pass through the locale the next day and entreating him to dissuade local whalers from catching her because she was on a pilgrimage to a shrine. If her wish were granted, she would let herself be caught on her way back from the pilgrimage. If her wish were not granted, evil spirits would befall the entire community and ruin it (see Table A.5).

These cautionary tales indicate that although the classical Japanese whalers did not have modern scientific knowledge, they had an instinctive understanding of the laws of ecology from their own experiences. They knew that pods of whales migrated from the north to the south in the fall and then migrated back to the north in the spring after the female whales gave birth in warm waters. They knew that catching a mother whale accompanied by a calf was an almost sure catch, albeit a very dangerous one: the mother whale would fight back hard in order to defend her calf and put whalers at greater risk, but she would eventually succumb to the attack because she would not leave her calf alone. This behavior is also observed when a pod of orcas (killer whales) attacks a mother whale and her calf.[6]

Nevertheless, classical Japanese whalers felt that they should refrain from catching mother whales and their calves. By likening a mother and calf's annual passage to the north to a human being's annual pilgrimage to a shrine, the folklore legends had taught them to forgo the short-term gain of catching such whales for the sake of the long-term preservation of the stock of whales. Therefore, if they caught a mother whale and her calf, despite the cautionary tales, they deeply feared punishment from the gods of the sea. They then attributed a bad future to the wrath of the spirit of the whale for violating the code of whaling concerning mother whales and their calves. Consequently, they sought for stronger atonement by creating graves and erecting monuments for the mother whale and her calf, as well as conducting elaborate funeral services and memorial services regularly afterward.[7]

In a final assessment, judging from the examination of scores of fishing communities throughout Japan that engaged in classical coastal whaling, it could be said that Japanese classical whalers mourned the deaths of whales and prayed for their souls so that the whales might attain the enlightenment of Buddha for a combination of three reasons: (1) empathy and sympathy for whales based on the Buddhist notion of benevolence for life and death; (2) atonement and remorse for killing whales based on the Buddhist proscription of killing living things; (3) and the fear of punishments from the gods or from the evil spirits deriving from superstition as told in folklore legends.

Demise of Classical Japanese Whaling

Classical Japanese whaling ended in the early twentieth century as the modern Western whalers had depleted the grounds surrounding Japan and whales stopped migrating off the Japanese coasts. Interestingly, Hashiura Yasuo holds the seclusionist *sakoku* policy of the Tokugawa shogunate government primary responsible for the demise of classical Japanese whaling at that time, instead of blaming it on American and European whaling operations in the Japan Grounds. The shogunate government issued a total of five decrees regarding the isolationist policy in 1633–1639 and enforced them until 1868, in order to monopolize foreign policy decision-making power and the interests of foreign trade (albeit limited) that was conducted in Nagasaki, as well as to suppress the influence of Portuguese and Spanish missionaries who were proselytizing Japanese into becoming Catholic (and then hoping to colonize Nagasaki and Kyūshū, as they did Macao and Manila).[8]

Hashiura Yasuo states that, as a corollary, the Tokugawa government not only banned construction of large ships, but also banned fishermen from sailing to the open sea, allowing them only to engage in coastal fishing. Going abroad or contacting foreigners was a crime punishable by death. Moreover, the shogunate government banned people, except for the samurai class, from possessing guns and even banned people from studying science or things Western. Thus, Hashiura argues that the seclusionist policy fatally delayed the modernization of Japan and obliged Japanese fishermen to engage in classical coastal whaling for three hundred years.[9]

Hashiura also asserts that, were it not for this oppressive policy, the Japanese would have developed their own modern whaling methods or adopted Western whaling methods much earlier than they did, and would have competed with Western whalers in the nineteenth century. In fact, in 1861, Nakahama Manjirō pleaded with the Tokugawa shogunate government to adopt the American methods using guns and bomb lances, and he actually tried this. However, it did not work because the Japanese did not have the knowhow due to the *sakoku* policy.[10]

In the absence of scientific knowledge and modern technology, the Japanese classical whalers blamed themselves for the disappearance of whales in the near seas (some of them thought that they had over-caught whales) and prayed vainly to the gods for the return of whales. It was not until after the overthrow of the Tokugawa shogunate government and the establishment of the Meiji government in 1868 that Oka Jūrō, a native of Nagato, Yamaguchi prefecture, sailed to Norway in order to learn the Norwegian whaling methods employing steam engine whaleships equipped with sophisticated bomb lance harpoons and guns invented by Svend Foyn in 1870. In July 1899, Oka and his colleague Yamada Tōsaku established the first modern Japanese whaling company, the Nippon Enyō-Gyōgyō Company (the prototype of the present Nippon Suisan) in Senzaki and its branch in Shimonoseki (see Chaps. 2 and 7).[11]

Distinction Between Classical and Modern Japanese Whaling

Thus, modern Japanese whaling developed during the mid-Meiji period (September 1868–July 1912). This was a drastic departure from classical whaling in many aspects. Here, Watanabe Hiroyuki importantly argues that modern Japanese whaling should not be regarded as an extension of classical Japanese whaling, but rather that the former developed almost as an antithesis to the latter. The distinctions between the two are not only in the methods and technology—classical Japanese whaling involved small-type coastal whaling (STCW) and only used manual equipment and manual rowboats, whereas modern whaling is large-type pelagic (open-sea, long-distance) whaling, employing advanced Norwegian whaling methods. They also differ in nature and rationale—classical whaling remained under the jurisdiction of the local provinces that issued licenses and financial credits to local whaling guilds, whereas modern whaling developed as a national industry under the protection and control of the Meiji government that launched the modernization and Westernization of Japan.[12]

In addition, Watanabe Hiroyuki argues that modern Japanese whaling was expansionist by nature in that the government mobilized colonial human resources—Korean labor—and constructed whaling bases in Korea (Japan annexed Korea in 1910). The Meiji government also encouraged whaling companies to construct processing (flensing) plants at locales in Japan without regard to local traditions and beliefs about whales. A prime example was the construction of a whale processing plant by the Tōyō Hogei Company in Same village, current Same-machi, Hachinohe, Aomori prefecture. The villagers opposed the construction because they had engaged in sardine fishing, and the whale processing plant polluted their fishing grounds. They also opposed it because they worshiped whales as the Ebisu God (whales enabled them to locate schools of sardines) and did not catch whales. On the morning of November 1, 1911, about eleven hundred fishermen in Same village burned the processing plant and attacked residences of company executives.[13]

In 1934, the Japanese ventured into pelagic whaling in the Antarctic Ocean, sending a fleet comprised of a large factory ship accompanied by several catchers. The Japanese government promoted pelagic whaling in the Antarctic because whale oil was an important export for earning foreign currency for Japan. Yet, whale meat was not a popular diet for Japanese at that time, partly due to inadequate methods for transporting frozen meat. During the time of classical coastal whaling, whale meat was a "random gift" from the gods of the sea. Even after Japan began modern whaling, whale meat remained a local delicacy in whaling communities and their vicinities. Meanwhile, with the rise of Japanese militarism, whaleships were turned into warships by the government, and all of the factory whaling ships were destroyed during World War II.[14]

After the defeat in World War II, the Japanese people suffered a serious food shortage, even worse than during wartime, and countless people starved to death. The General Headquarters of the Supreme Commander of the Allied Powers (SCAP-GHQ), composed of US occupation forces administering the Japanese postwar government, sanctioned the restoration of pelagic whaling in August 1946 in order to alleviate the food crisis and to secure a stable supply of animal protein for the Japanese. In this context, Japanese whaling companies resumed pelagic operations in the Antarctic

Ocean in full swing. Consequently, through the enactment of the School Lunch Act in 1954, whale meat appeared on lunch tables of public schools in order to offset the nutritional deficiencies of Japanese children.[15]

In turn, the introduction of whale meat in school lunch menus made whale eating "part of the nationwide food culture" in Japan for the first time in its history, not by popular demand, but by government policy. Thus, it took the defeat in World War II for the local whale-eating culture to go beyond its geographic and subcultural boundaries and for the Japanese people to eat canned cooked whale meat, as people eat canned tuna today.[16]

However, Jun Morikawa is quick to point out the fact that this phenomenon lasted less than two decades during the postwar reconstruction period from the late 1940s to the early 1960s. As Japan's economy recovered and people could afford more expensive meat (such as beef and pork), whale meat disappeared from their dinner tables, as well as from school lunch tables. Thus, as fast as the supply of whale meat rose in the late 1940s, its demand dropped sharply in the early 1960s. The fact was that whale meat has never really been a popular meat for the Japanese at large by choice, but was only a "substitute meat." Morikawa states that this is the truth about the so-called "a long history of whale eating culture in Japan."[17]

Morikawa further argues that it is a recently created myth used to justify Japanese whaling in the Antarctic Ocean in the face of international criticism. With the imposition of a moratorium on commercial whaling in the Antarctic, whale meat has reverted to being a local delicacy of certain fishing communities in Japan, where STCW is conducted legally today in accordance with the International Whaling Commission (IWC), and so has become beyond the reach of ordinary people. Against this current, however, half a century after the legislation of the School Lunch Act, the conservative Koizumi cabinet, ruled by the Liberal Democratic Party (LDP), legislated the Basic Law on Food Education (*shokuiku*) in 2005. Since then, the Japanese government has incomprehensively been encouraging local boards of education to reintroduce whale meat to lunch menus.[18]

Small-Type Coastal Whaling (STCW)

Overall, from a wider historical perspective, all of the four major classical coastal whaling bases in Japan stopped operations during the Meiji period, with the one exception of Taiji, Wakayama prefecture. Also, with the exception of Taiji in Wakayama prefecture, none of the five fishing communities in Japan that engage in STCW today—Abashiri and Hakodate, both in Hokkaidō, Ayukawa in

Miyagi prefecture, and Wada in Chiba prefecture—were major classical whaling bases. The IWC has not banned STCW of smaller species of whales that are not endangered or threatened, such as Baird's beaked whales and short-finned pilot whales. Japanese STCW is operated under strict regulations and quotas determined by the Japanese government in accordance with the guidelines set forth by the IWC.[19]

IWC Moratorium and ICJ Ruling

Regarding the moratorium on commercial whaling in the Antarctic Ocean adopted by the IWC (its 2002 annual general meeting was held in May in Shimonoseki, Yamaguchi prefecture, which was a pelagic whaling base), a former official at Japan's Fisheries Agency and an acting Japanese representative to the IWC, Komatsu Masayuki, criticized the Japanese government's stance. He was then abruptly demoted by the government in April 2005. Komatsu argued that so long as the Japanese government wanted to continue pelagic operations in the Antarctic, it should have opted out of the moratorium clause of the International Convention for the Regulation of Whaling (ICRW) and engaged in operations on its own outside the framework of the IWC, as Iceland and Norway did (Iceland later adopted the clause). Or Japan even could have withdrawn from the international regime entirely, as Canada did. Instead, the Japanese government compromised on its position and withdrew its objection to the moratorium clause and accepted it.[20]

Nevertheless, Japan soon afterwards resumed pelagic operations in the Antarctic under the "scientific research" provision of the ICRW. This led to international accusations against Japan as being engaged in commercial whaling under the pretext of "scientific research." The controversy escalated to the extent that the Australian government filed a case in the International Court of Justice (ICJ) in May 2010, which resulted in a ruling against Japan in March 2014.[21]

In hindsight, Komatsu feels that had the Japanese government opted out of the moratorium clause of the ICRW, it might not have been "singled out" by anti-whaling advocacy groups as an "easy target," and the case might not have been brought by the Australian government in the ICJ. Yet, this is easier said than done. The Japanese government was under pressure from anti-whaling members of the IWC, such as the United States, to accept the moratorium. Japan succumbed to it, as has often been the case for Japanese policy making under *gaiatsu* (foreign pressure).[22]

Monuments Created by Modern Japanese Whalers

With all things considered regarding the disparities between classical and modern Japanese whaling, including nature, objective, and rationale, there is one thing in common between the two. This concerns the ways in which the modern whalers have understood their profession and dealt with the deaths of the whales they caught. The commonalities between classical and modern whaling is prominent in the whale monuments that modern Japanese whalers have created. As with their classical predecessors, they wanted to atone for their deeds and to pray for the souls of the whales that they caught. For example, executives and harpooners of Japanese whaling companies erected brand new, respectable monuments in Hakodate, Ayukawa, Shirahama (Minami-Bōsō), Taiji, Kure, Muroto, and Shimonoseki (see Chaps. 3, 4, 5, 6, 7, 8 and 9; Table A.1).

Modern whalers and fishermen have also conducted memorial services for whales annually on seasonal Buddhist memorial days. Such cities and towns as Ayukawa, Taiji, Nagato, Ōtomari, Usuki, Ebisu-ura in Ashibe-chō on Iki Island, and Shin-Uonome on the Gotō Islands have observed this practice to this day. In this sense, although the rationale and methods of whaling have changed from classical whaling to modern whaling, the compassionate hearts of the classical Japanese fishermen and whalers have been passed down to their modern counterparts (see Chaps. 3, 4, 5, 6, 7, 8 and 9; Table A.2).

Universal Applicability of Japanese Culture of Mourning the Deaths of Whales

In closing, the Japanese culture of mourning the deaths of whales should not be misconstrued as an ethnocentric notion. This culture is based on the syncretism of Shintoism and Buddhism and subscribes to a fundamental reverence for life and benevolence for life and death. This is not a unique concept that could be only understood and practiced in the cultural background of Asia, as Albert Schweitzer embraced the notion of reverence for all forms of life as the guiding moral principle for all human beings. Also, the holistic way in which many Japanese have grasped and treated nature can serve as a universal guidepost for its conservation and the management and sustainable use of natural resources today.[23]

The Japanese culture of mourning the deaths of whales has a universal applicability because it is a manifestation of veneration of nature. Many of its aspects—receiving whales as blessings of the gods of the sea, consuming them with appreciation, utilizing every usable part without wasting anything, and taking care of their deaths and providing for their "afterlives"—are all relevant to contemporary issues of conservation of nature and natural resources, as well as protection of wildlife. A proponent for "creative conservation" at Wakayama University, Kumi Kato, who visited the former classical whaling base in Kayoi, Nagato, Yamaguchi prefecture, argues in her article written in English that the whaling culture of Kayoi presents a case for sustainable management of natural resources.[24]

Now that the age of the pelagic whaling in the Antarctic is over for Japan, only a few small communities are engaged in STCW. The local whalers this author met in Ayukawa and Taiji were fishermen who made a living primarily by fishing and occasionally by whaling. The international scholars workshop on STCW concludes that Japanese STCW has elements of "subsistence whaling" as well as of commercial whaling. Accordingly, for these communities it is important to try to find a reasonable balance between the stock of unthreatened whale species and an annual quota of STCW for them.[25]

In conclusion, contemporary Japanese STCW whalers must firmly embrace the hearts of their classical predecessors who mourned the deaths of whales as if they were human beings and treated them as such. They should place an utmost priority on the protection and preservation of whales in this age in which many species of whales are endangered or threatened. It is a mandate for Japanese policy makers and those who are involved in contemporary whaling to identify rational annual quotas for Japanese STCW. In so doing, they must work closely with international organizations, such as the IWC, and take every measure to ensure sustainable growth of the populations of the various whale species.

It is this author's hope that mankind will take more effective measures to protect wildlife—both fauna and flora—preserve their habitats on land and sea, and ensure sustainable use of natural resources. The Japanese should carry on the mentality of their ancestors, who extended sympathy to the animals and plants that they consumed, as blessings of nature, and should make every effort toward this universal goal of mankind. The hearts of the Japanese fishermen-whalers who mourned the deaths of whales and prayed for their souls will live forever in the whale graves and monuments they created. In turn, these monuments will forever watch over whales that migrate near Japan in the Pacific Ocean and in the Sea of Japan, and also serve as lighthouses to guide mankind toward the universal goal of conservation of nature and natural resources.

NOTES

1. Kawamura Shigeyoshi, author interview, August 6, 2010.
2. Ibid.
3. Tsunoda Fusako, "Ōmi-jima: Nisō no iru chi no hate" (Ōmi Island: The Edge of the World Where Nuns Live), *Tabi*, September 1969, 72.
4. Hosokawa Takao, "Kujira ni 'kansha to ikei' no nen" (Feelings of "Appreciation and Respect" for Whales), *Mainichi Shimbun*, October 26, 2010.
5. Boreal Institute for Northern Studies, ed., *Small-Type Coastal Whaling in Japan: Report of an International Workshop*, Edmonton, Alberta: Boreal Institute for Northern Studies, the University of Alberta, 1988, 53–57, 61–62.
6. Ibid., 62; Kon'no Tsutomu, *Kaneko Misuzu futatabi* (Revisiting Kaneko Misuzu), Tokyo: Shōgakukan, 2007, pbk., 2011, 214–215.
7. Ibid. (both).
8. Kumano Taiji-ura hogei-shi hensan-iinkai, ed., *Kumano no Taiji: Kujira ni idomu machi* (Taiji, Kumano: Town That Challenges Whales), Tokyo: Heibonsha, 1965, 24–26.
9. Ibid.
10. Ibid.; Boreal Institute for Northern Studies, 12.
11. Ibid. (both).
12. Watanabe Hiroyuki, *Hogei-mondai no rekishi shakaigaku: Kingendai Nihon ni okeru kujira to ningen* (Historical Sociology of Whaling Problems: Whales and People in Modern Japan), Tokyo: Tōshindō, 2006, 17–56.
13. Ibid., 32–33, 57–58, 72; Jun Morikawa, *Whaling in Japan: Power, Politics, and Diplomacy*, New York: Columbia University Press, 2009, 21–23.
14. Morikawa, 23–28.
15. Ibid., 28–30, Boreal Institute for Northern Studies, 14–15, 67–70.
16. Ibid. (both).
17. Morikawa 30–33; Boreal Institute for Northern Studies, 94.
18. Ibid.
19. Boreal Institute for Northern Studies, 18–25.
20. Komatsu Masayuki, *Sekai kujira sensō* (World War on Whales), Tokyo: PHP kenkyūjo, 2010, 197–218; "Komatsu Masayuki-shi wa Beikoku no atsuryoku de kōshō kara hazusareta" (Mr. Komatsu Masayuki Removed from Negotiations by U.S. Pressure), October 22, 2005, http://blog.livedoor.jp/standupjoe1/archives/50121502.html
21. Ibid. (both).
22. Ibid. (both); Isao Miyaoka, "State-Society Relations Under Foreign Pressure: Two Contrasting Cases in the Japanese Fisheries Policy," http://www.dijtokyo.org/doc/dij-jb_11-miyaoka.pdf, downloaded February 24, 2016. For details of *gaiatsu* (foreign pressure) politics, see Kent E. Calder,

"Japanese Foreign Economic Policy Formation: Explaining the Reactive State," *World Politics*, Vol. 40, No. 4, July 1988, 517–541.

23. Arne Kalland, "Holism and Sustainability: Lessons from Japan," *Worldviews: Global Religions, Culture, and Ecology* (2002), Vol. 6, No. 2, 145–158. For detail, see Patricia J. Asquith and Arne Kalland, *Japanese Images of Nature: Cultural Perspectives*, Richmond, Surrey, UK: Curzon Press, 1997.
24. Ibid.; Kumi Kato, "Prayers for the Whales: Spirituality and Ethics of a Former Whaling Community—Intangible Cultural Heritage for Sustainability," *International Journal of Cultural Property* (2007), No. 14, 283–313.
25. Boreal Institute for Northern Studies, 79–84.

Appendix 1

Table A.1 Whale graves and related monuments in Japan

Location	*Year*	*Shrine/temple*	*Description*
Hokkaidō Region			
1. Yayoi, Hakodate	1957		Memorial cenotaph for whales
Tōhoku Region			
Iwate Prefecture			
2. Ryōishi	1858		Memorial stone for Steller sea lions
3. Hirota	1905		Whale grave and gravestone
4. Hirota	1704–1711	Kan'non-dō	Wooden statue of Thousand-armed Goddess of Mercy for Whales
5. Hirota	1704–1711?	Kan'non-dō	Gold-painted statue of Thousand-armed Goddess of Mercy for Whale
Miyagi Prefecture			
6. Karakuwa	1810	Osaki Shrine	Whale grave
7. Karakuwa	1835	Osaki Shrine	Whale grave
8. Karakuwa	undated	Osaki Shrine	Whale grave
9. Osaki	1875	Osaki Shrine	Whale grave
10. Minami-Sanriku	1746–1747		Memorial stupa for whales
11. Ayukawa		Kan'non Temple	Grand memorial tablets for whales
12. Ayukawa	1932	Kan'non Temple	Memorial stupa for whales
13. Ayukawa	1933	Kan'non Temple	"Memorial Stupa for One Thousand Whales"

(*continued*)

M. Itoh, *The Japanese Culture of Mourning Whales*,
https://doi.org/10.1007/978-981-10-6671-9

Table A.1 (continued)

Location	*Year*	*Shrine/temple*	*Description*
Akita Prefecture			
14. Funakawa-Minato	1889	"Whale School"	Memorial stone for whale
Kantō Region			
Chiba Prefecture			
15. Chōshi	undated	Kawaguchi Shrine	Whale grave
16. Chikura	1896	Chōshō Temple	Whale grave with decorative five-tiered memorial stupa (*hōkyōin-tō*)
17. Shirahama	1871 (1916)		two gravestones for mother whale and fetus
18. Shirahama	1947		Cube memorial stupa for whales
19. Katsuyama	undated		Whale grave for fetus
20. Katsuyama	1612–	Benzaiten Shrine	More than 120 whale gravestones
21. Urayasu	1875	Inari Shrine	Four memorial stones for a whale
22. Urayasu	1892	Kezō-in Temple	Whale grave and memorial stupa
Tokyo Prefecture			
23. Higashi-Shinagawa	1798	Kagata Shrine	Whale grave and memorial stone
24. Higashi-Shinagawa	1969	Kagata Shrine	Memorial stone for whale
25. Samezu	1851	Samezu Hachiman Shrine	Whale grave
Kanagawa Prefecture			
26. Misaki	1834	Nishihama Jizō-dō Temple	Whale grave
Chūbu and Hokuriku Regions			
Shizuoka Prefecture			
27. Kawana	1923	Mishima Shrine	Memorial stone for dolphins
28. Kawana	1977	Kaizō Temple	Memorial stupa for dolphins
29. Inatori	1828		Memorial stupa for dolphins
30. Arari-ura	1882		Memorial stone for dolphins
31. Arari-ura	1935		Memorial stupa for dolphins
32. Arari-ura	1949		Memorial stupa for dolphins
33. Toi	1959	Ōyabu Benten Shrine	Memorial stone for dolphins

(*continued*)

Table A.1 (continued)

Location	*Year*	*Shrine/temple*	*Description*
Niigata Prefecture			
34. Maze	1934	Kaiun Temple	Whale grave
35. Nozumi	undated	Saishō Temple	Sacred skull of whale
36. Matsuzawa	1849		Whale grave
37. Ōwada	1897	Kan'non-dō	Whale grave
38. Nishiyama	1956	Shinzō-in Temple	Memorial stupa for whales
39. Kashiwazaki	1910	Miyagawa Shrine	Memorial stone for a whale
40. Katano'o	1861 & 1867		Whale grave, jawbone gravemarker, memorial stupa for prayer of one million chanting, and memorial prayer stone
41. Shiidomari	1888	Gansei Temple	Whale grave
42. Haniu	1881/1882		Whale grave and memorial stupa
43. Bentenzaki	1985		Whale grave and memorial stupa
Fukui Prefecture			
44. Arawa	1812		Whale Grave and memorial stone
45. Aoto no Irie	1918		Whale grave and memorial stone
Kansai-Kinki Region			
Mie Prefecture			
46. Futami	1897		Whale grave and memorial stone
47. Futami	1903		Whale grave and memorial stone
48. Ōsatsu	1989		Memorial stone for whales
49. Daiō		Nakiri Shrine	Repository for Hiruko God
	circa 1700		"Whale Stone"
50. Shiroura	1759	Jōrin Temple	Grave and memorial stupa for pregnant whale
51. Shiroura	undated	Jōrin Temple	Two memorial tablets for pregnant whale
52. Shiroura	1957	Jōrin Temple	Created Kaikō Shrine and new memorial stone for pregnant whale at 200-year memorial anniversary
53. Owase	1841	Fusai Temple	Memorial stupa for whales
54. Nigishima		1671	"Memorial Stupa for Thirty-Three Whales"
55. Nigishima			Many memorial stones for whales
56. Kinomoto	1880		Memorial stone for whale

(*continued*)

Table A.1 (continued)

Location	*Year*	*Shrine/temple*	*Description*
Wakayama Prefecture			
57. Taiji	1768	Tōmyō Temple	Whale grave and memorial stupa
58. Taiji	circa 1800	Tōmyō Temple	Memorial prayer stone for whales
59. Taiji	undated	Tōmyō Temple	Statue of Goddess of Mercy for Fish
60. Taiji	1986	Tōmyō Temple	New Stone Statue of Guardian Deity of Unborn Children for Whales
61. Taiji	1979	Kandori-zaki Park	Memorial cenotaph for whales
62. Kushimoto	1940		Memorial cenotaph for whales
Osaka Prefecture			
63. Higashi-Yodogawa	undated	Zuikō Temple	"Snow Whale Bridge"
Kyoto Prefecture			
64. Ine	1808	Ebisu Shrine	Graves and three memorial stones for calf, fetus, and mother whale
Chūgoku Region			
Shimane Prefecture			
65. Kizuki-Kita	undated	An'yō Temple	Whale grave
Yamaguchi Prefecture			
66. Mishima	1711–1716		A thousand whale graves (two dozen of them remain)
67. Kayoi	1692	Seigetsu-an	"Ōmijima Whale Grave" for Fetuses
68. Kayoi	1692	Kōgan Temple	Grand memorial tablet for whales
69. Kayoi	1692–1908?	Kōgan Temple	Four-volume death register recording more than a thousand afterlife names of whales
70. Kayoi	1863	Kōgan Temple	"Stone Statue of Guardian Deity of Unborn Children for Souls of Whales"
71. Kayoi	undated	Whale Museum	Wooden statue of Goddess of Mercy for Souls of Fish, riding on mother whale with calf
72. Ōhibi		Hōsen-an nunnery	Death register including whales
73. Senzaki		Gokuraku Temple	Grand memorial tablet for whales
74. Senzaki	1759	Fumon Temple	Memorial stone for fish (including whales)

(*continued*)

Table A.1 (continued)

Location	*Year*	*Shrine/temple*	*Description*
75. Kiwado		Kaigan Temple	Memorial stone for whales (weathered away)
76. Tsuō-Tateishi			Stone statue of Goddess of Mercy
77. Kawajiri			Whale graves (buried nasal bone) and gravestones
78. Kawajiri	1897		Wooden statue of Goddess of Mercy for Whales
79. Kawajiri	1961		Memorial stone for souls of whales
80. Shimonoseki	2002		Memorial cenotaph for whales
Okayama Prefecture			
81. Ushimado	1884		Whale grave and gravestone
Hiroshima Prefecture			
82. In'noshima		Jigen Temple Taichō-in	"Statue of Goddess of Mercy for Whales"
83. Kure	1983		Whale grave and memorial stone
Shikoku Region			
Kōchi Prefecture			
84. Ukitsu		Chūdō Temple	Many memorial stupas for whales
85. Ukitsu	1837	Chūdō Temple	Grand bronze hanging bell
86. Ukitsu	1840	Chūdō Temple	Three grand memorial tablets for whales
87. Moto'otsu	1960	Kongōchō Temple	"Memorial Cenotaph for Souls of Eight Thousand Whales"
88. Moto'otsu		Kongōchō Temple	Grand bronze hanging bell
89. Moto'otsu		Kongōchō Temple	Memorial Museum for Whales
90. Moto'otsu	1967	Kongōchō Temple	"Memorial Cenotaph for Souls of Eight Thousand Whales"
91. Kubotsu	1812	Kaizō-in Temple	"Stone Statue of Guardian Deity of Unborn Children for Souls of Whales"
Ehime Prefecture			
92. Uchidomari	1882	Sairin Temple	Whale grave and gravestone
93. Ōhama	1900	Sairin Temple	Whale grave and gravestone
94. Miji-no-shima	1806–1846		Whale grave and memorial stone
95. Futanarabi-jima	1834		Whale grave and memorial stone
96. Amayadori-jima	1975		Whale grave and memorial stone
97. Koyanoura, Yusu	1977		Memorial stone for whales
98. Banshō, Yusu	1986		Whale grave and memorial stone

(continued)

Table A.1 (continued)

Location	*Year*	*Shrine/temple*	*Description*
99. Takasuke	1892	Mankichi Temple	Whale grave, conferred afterlife name on whale and recorded it in death register
100. Mizutani	1948		Whale grave and gravestone
101. Akehama	1837		"Hayanote Whale Grave" with memorial stone
102. Akehama		Kongō Temple	Memorial tablet for whale, conferred afterlife name on whale and recorded it in death register
103. Kusunoura	1870		"Kusunoura Whale Grave" with memorial stone
104. Miyanotsu	1907		"Whale Grave at Komochi Rock" with memorial stone
105. Mitsukue	1809		Whale grave and gravestone, with afterlife name engraved on gravestone
106. Kojima	1879		Whale grave and gravestone, with afterlife name engraved on gravestone
107. Kawanohama	1887		Whale grave and memorial tower
108. Natori	1929		Whale grave and memorial stone
109. Kushi	1882		Whale grave and two memorial stones (weathered away), erected new memorial tower
110. Kamihama	1910	Minato Shrine	Whale grave and memorial tower
111. Ozaki	1910		Memorial tower
Kyūshū Region			
Fukuoka Prefecture			
112. Kutsuo	1903		Whale grave and gravestone
113. Hakozaki	1888		Whale grave and gravestone, transferred to Amiya Tenjin Shrine in 1994
Ōita Prefecture			
114. Bungo-Takada	1843		Whale grave and gravestone
115. Isshakunoya	1855		Whale grave and gravestone
116. Isshakunoya	1888		Whale grave and gravestone
117. Isshakunoya		Kyōson Temple	Death register with afterlife name of whale
118. Sashiu, Usuki	1886		Whale grave and gravestone
119. Sashiu, Usuki		Ōhashi Temple	Death register with afterlife name of whale
120. Ōhama, Usuki	1894		Whale grave and gravestone, conferred afterlife name
121. Ōhama, Usuki	1944		Whale grave and gravestone

(*continued*)

Table A.1 (continued)

Location	*Year*	*Shrine/temple*	*Description*
122. Ōtomari, Usuki	1870	Ōhashi Temple	Whale grave and gravestone
123. Ōtomari, Usuki			Memorial tablet for whale
124. Ōtomari, Usuki			Death register with afterlife name of whale
125. Ōtomari, Usuki	1871		Erected grand stupa at first anniversary memorial day
126. Itachiya, Usuki			"Stone Statue of Guardian Deity of Unborn Children for Whales"
127. Nakatsu-ura, Usuki	1882	Ebisu Shrine	Whale grave and gravestone
128. Azamui, Kamiura	1888		Whale grave and gravestone
129. Azamui, Kamiura	1907		Whale grave and gravestone
Miyazaki Prefecture			
130. Aburatsu		Shōgyō Temple	Whale grave and gravestone, transferred to Tobashira Park
Saga Prefecture			
131. Kimizuka, Yobuko	1748		Whale grave and gravestone
132. Yobuko	1714	Ryūshō-in Temple	Memorial stupa for whales, with memorial wooden tablets with afterlife names of whales
133. Yobuko	1813	Ryūshō-in Temple	Five-tiered memorial stupa for whales with stone reliefs of Six Guardian Deities for Unborn Children
134. Yobuko	1831	Ryūshō-in Temple	"Memorial Stupa for One Thousand Whales"
135. Ogawa Island	1863	Kan'non Temple	Memorial stupa for whales with relief of Guardian Deity for Unborn Children
136. Karatsu		Kōju-in Temple	Memorial stone for whales
Nagasaki Prefecture			
137. Isahaya	1885		Whale grave and memorial stone
138. Ashibe, Iki Island			Grave and gravestone for mother whale and her fetus
139. Ashibe, Iki Island	1717		Whale grave and memorial stone (*kasa-tōba*)

(*continued*)

Table A.1 (continued)

Location	*Year*	*Shrine/temple*	*Description*
140. Numatsu, Iki Island		Kaihō Temple	Hanging bell with a relief of whale (stolen by pirates)
141. Katsumoto, Iki Island	2000	Nagatori residence	Memorial stone for whales
142. Tatsunoshima Island			Memorial cenotaph for the souls of dolphins
143. Iwanoue, Hirado	1695	Saikyō Temple	Whale grave and memorial stupa
144. Iwanoue, Hirado	1938	Saikyō Temple	Whale grave and five-tiered memorial stupa (*gorin-tō*)
145. Mae-Tsuyoshi, Hirado	1739	Chōsen Temple	Grand five-tiered memorial stupa for whales with statues of Abundant Treasures Buddha and Amida Buddha (*gorin-tō*)
146. Azuchi-Ō shima Island	1692	Shinkyō Temple	Whale grave and gravestone
147. Taira, Uku Island		Tōkō Temple	Whale grave and gravestone
148. Taira	1924	Tōkō Temple	"Grave for One Thousand Whales"
149. Ojika Island	1695		Whale grave and memorial stone
150. Aokata		Fukuju-an Temple	Whale grave and pebbles for prayer of copying down sutra with one character on one stone and burying them
151. Arikawa	1712		Grave and "Memorial Stone for 1312 Whales"
152. Arikawa			New memorial stone
153. Uonome	1696		Grave and memorial stone for 106 whales
154. Uonome	1774		Grave and memorial stone for one hundred whales
155. Kojima			Grave, gravestone, and memorial stone
156. Kurose	1690		Grave and gravestone

Notes: The number of each entry corresponds to that in the Map

Sources: Yoshihara Tomokichi, "Kujira no haka" (Graves of Whales), in Tanigawa Ken'ichi, ed., *Nihon minzoku bunka shiryō shūsei* (Collection of Documents on Japanese Folk Culture), Vol. 18, Tokyo: San'ichi-shobō, 1997, 409–472; "Kujira bunka: Kujira o tomuratta kujira-baka · kujira-zuka nado" (Whale Culture: Whale Graves and Monuments to Mourn the Deaths of Whales), http://www.geocities.co.jp/NatureLand-Sky/3011/kujirahaka.html, accessed November 14, 2015

Appendix 2

Table A.2 Memorial rites for whales and other related intangible cultural properties

Location	*Shrine/temple*	*Description*
Iwate Prefecture		
Hirota	Jion Temple	Annual memorial service for Thousand-armed Goddess of Mercy for Whales (Autumn)
Miyagi Prefecture		
Karakuwa	Osaki Shrine	Annual memorial service (January)
Ayukawa	Kan'non Temple	Annual *segaki* memorial service (August)
Niigata Prefecture		
Katano'o	Jizō-in Temple	Offered grand funeral service (prayer of chanting sutra one million times), conferred afterlife name on whale, and recorded it in death register (1861), and conducted seventh-year memorial service (1867)
Shiidomari	Gansei Temple	Conferred afterlife name on whale (1888)
Mie Prefecture		
Kusu-chō		Annual whale boat memorial service
Yokkaichi		Annual whale boat memorial service
Miyama		Annual whale boat memorial service
Shiroura		Annual memorial service for a pregnant whale
Wakayama Prefecture		
Taiji	Tōmyō Temple	Offered prayer of copying down sutra on stones and burying them

(*continued*)

M. Itoh, *The Japanese Culture of Mourning Whales*,
https://doi.org/10.1007/978-981-10-6671-9

Table A.2 (continued)

Location	*Shrine/temple*	*Description*
Taiji	Tōmyō Temple and Junshin Temple	Annual memorial service (April)
Taiji	Tōmyō Temple and Junshin Temple	Annual *segaki* memorial service (August)
Taiji	Asuka Shrine	Archival documents on memorial
Taiji	Asuka Shrine	Rites for fetuses of whales, originally preserved at Enkai-in Temple
Yamaguchi Prefecture		
Kayoi	Kōgan Temple	Annual memorial service for whales (April)
Ōhibi	Hōsen-an nunnery	Daily prayers includes those for souls of whales
Senzaki	Gokuraku Temple and Enkyū Temple	Take turns in conducting annual memorial services for whales (April)
Senzaki	Jōgan Temple, Saikaku Temple, and Henshō Temple	Take turns in conducting annual memorial services for whales (April)
Misumi	Seifuku Temple	Annual memorial service for whales
Kawajiri		Annual memorial service for whales (spring)
Ehime Prefecture		
Akehama	Kongō Temple	Annual memorial service for a whale
Kawanohama		Offered prayer of copying down Mahayana Buddhism Sutra with one character on one stone and burying them (1887)
Kamihama	Minato Shrine	Offered prayer of copying down Mahayana Buddhism Sutra with one character on one stone and burying them (1910)
Ozaki		Offered prayer of copying down Mahayana Buddhism Sutra with one character on one stone and burying them (1910)
Ōita Prefecture		
Ōdomari, Usuki		Ōhashi Temple Grand funeral service and *segaki* ritual, offered prayer of chanting sutra one million times (1870)
Ōdomari, Usuki		Ōhashi Temple Annual memorial services and 100-year anniversary memorial service (1969)
Saga Prefecture		
Yobuko	Ryūshō-in Temple	Annual memorial services and rituals of sending off souls of whales to the sea (Bon, week of August 15)
Nagasaki Prefecture		
Ashibe, Iki Island		Offered chanting of one thousand copies of Mahayana Buddhism Sutra (1717) and conducted annual memorial services for whales

(*continued*)

Table A.2 (continued)

Location	*Shrine/temple*	*Description*
Taira, Uku Island	Tōkō Temple	Annual memorial service for whales (spring)
Aokata, Nakadōri Island	Fukuju-an Temple	Offered prayer of copying down sutra with one character on one stone and burying them
Uonome, Nakadōri Island		Offered prayer of copying down sutra with one character on one stone and burying them (1696)

Sources: Yoshihara Tomokichi, "Kujira no haka" (Graves of Whales), in Tanigawa Ken'ichi, ed., *Nihon minzoku bunka shiryō shūsei* (Collection of Documents on Japanese Folk Culture), Vol. 18, Tokyo: San'ichi-shobō, 1997, 409–472; "Kujira bunka: Kujira o tomuratta kujira-baka · kujira-zuka nado" (Whale Culture: Whale Graves and Monuments to Mourn the Deaths of Whales), http://www.geocities.co.jp/NatureLand-Sky/3011/kujirahaka.html, accessed November 14, 2015

Appendix 3

Table A.3 Selected monuments dedicated to Ebisu god and other tangible cultural properties related to whaling

Location	*Year*	*Shrine*	*Description*
Miyagi Prefecture			
Tomari	1891	Kakura-Myōjin Shrine	Whale *ema*
Ajishima			Memorial stone for Grand God of Whales
Chiba Prefecture			
Katsuyama		Benzaiten Shrine	Monument of Benzaiten for whales
Katsuyama	1836	Kachiyama Shrine	Whale *ema*
Shirahama		Oceanic Art Museum	Picture scroll of classical whaling
Ishikawa Prefecture			
Unzu	1848–1855		Drawing of net method of whaling
Unzu			"Picture of Lord Tō Watching Whale Catching"
Mie Prefecture			
Ōsatsu			Monument for original site of Eleven-faced Goddess of Mercy
Owase	1764	Kuki Shrine	Donation of pair of stone lanterns

(*continued*)

M. Itoh, *The Japanese Culture of Mourning Whales*,
https://doi.org/10.1007/978-981-10-6671-9

Table A.3 (continued)

Location	*Year*	*Shrine*	*Description*
Wakayama Prefecture			
Taiji	before 1688	Ebisu Shrine	Whale jawbone gate
Taiji			*Kujira-bune emaki* ("Picture Scroll of Whale Boats")
Taiji			Pieces of whale boats with designs from pictures of Pure Land of Paradise
Kushimoto	1908	Kushimoto Jitsugyō School	Whale jawbone gate
Shimane Prefecture			
Setogashima	1854	Itsukushima Shrine	Whale *ema*
Yamaguchi Prefecture			
Senzaki		Yasaka Shrine	*Kujira hokaku-ezu* ("Picture of Whaling")
Kayoi		Whale Museum	Equipment for classical whaling
Nagato			Color illustrations and pictures of whaling
Hiroshima Prefecture			
In'noshima	1899	Hachiman Shrine	Whale *ema*
Kagawa Prefecture			
Kotohira		Kotohira Shrine	Whale *ema*
Kōchi Prefecture			
Tosa-Saga	1652	Kashima Shrine	*Waniguchi* hanging bell
Ehime Prefecture			
Kawanoe	1864	Hachiman Shrine	Whale *ema* plaque made of whale bone
Ōita Prefecture			
Bungo-Takada	1888	Wakamiya Hachiman Shrine	Whale *ema* (lost)
Saga Prefecture			
Yobuko		Atago Shrine	Stone lanterns with *komainu* standing on hind legs
Yobuko		Hachiman Shrine	Stone statue of Ebisu God
Yobuko		Hachiman Shrine	*Waniguchi* hanging bell
Yobuko	1840	Shūhōtei Riyū	*Ogawa-jima geigei gassen* (Chronicle of "Whale Battles at Ogawa Island")

(*continued*)

Table A.3 (continued)

Location	*Year*	*Shrine*	*Description*
Yobuko			"Picture of Memorial Service for Whales at Ryūshō Zen Temple" in *Ogawa-jima geigei gassen* (1840)
Yobuko			"Picture of Memorial Rite of Transferring Souls of Whales by Sending Them off to the Sea" in *Ogawa-jima geigei gassen* (1840)
Ogawa Island		Tajima Shrine	Stone *torii*
Kashiwa Island	1895		"Stone Statue of Ebisu God Riding on Whale"
Nagoya	1856	Furusato Shrine	Stone statue of Ebisu God
Taniga Island	1921		Stone statue of Ebisu God and Great God of Dolphins
Taniga Island	1943		Stone statue of Ebisu God and Great God of Dolphins
Nagasaki Prefecture			
Ikitsuki Island		Misaki Shrine	Statue of Ebisu God with Whale
Ikitsuki Island		Ikitsuki Township Museum	Two-volume *Isana-tori e-kotoba* ("Illustrations of Whaling")
Arikawa, Nakadōri Island Kaidō Shrine			Whale jawbone gate

Notes: This table lists only shrines that are known to have whale monuments and objects in dedication to the Ebisu God

Sources: Yoshihara Tomokichi, "Kujira no haka" (Graves of Whales), in Tanigawa Ken'ichi, ed., *Nihon minzoku bunka shiryō shūsei* (Collection of Documents on Japanese Folk Culture), Vol. 18, Tokyo: San'ichi-shobō, 1997, 409–472; "Nihon no kujira bunka: Kujira to shinkō" (Japanese Whale Culture: Whales and Beliefs), http://www.geocities.co.jp/NatureLand-Sky/3011/kujira-to-sinkou.html, accessed November 23, 2015

Appendix 4

Table A.4 Selected intangible cultural properties dedicated to whale god

Location	*Description*
Hokkaidō	
Shiraoi-chō	Whale dance and whale festival
Hamanaka	*Kiritappu-hama drum*
Miyagi Prefecture	
Ayukawa	Kumano Shrine summer and autumn festivals
Ayukawa	Whale festival
Chiba prefecture	
Katsuyama	Whale festival
Mie Prefecture	
Suzuka	Ten'no festival and whale boat song
Yokkaichi	Whale festival
Ōsatsu	Ten'no whale festival and dance
Owase	Whale festival
Shingū	Whale dance
Wakayama Prefecture	
Taiji	Whale boat race
Taiji	Hometown festival
Osaka Prefecture	
Dōjima, Osaka	Whale dance
Yamaguchi Prefecture	
Nagato	Kayoi whale song

(*continued*)

M. Itoh, *The Japanese Culture of Mourning Whales*,
https://doi.org/10.1007/978-981-10-6671-9

Table A.4 (continued)

Location	*Description*
Ehime Prefecture	
Kashiwazaki	No-fishing-day Festival (September 6)
Akehama	150-year anniversary whale dance
Miyazaki Prefecture	
Aburatsu	Tradition of making rice cake in shape of whale during Bon week
Saga Prefecture	
Itsuki	Whale song recital
Nagasaki Prefecture	
Yobuko	Whale processing song
Ashibe, Iki Island	Whale song
Taira, Uku Island	Ebisu Shrine Koshiki Festival (November)
Fukue Island	Whale song dedicated to "Monkurō Whale" (January 16)

Sources: Yoshihara Tomokichi, "Kujira no haka" (Graves of Whales), in Tanigawa Ken'ichi, ed., *Nihon minzoku bunka shiryō shūsei* (Collection of Documents on Japanese Folk Culture), Vol. 18, Tokyo: San'ichi-shobō, 1997, 409–472; "Nihon no kujira bunka: Kujira to shinkō" (Japanese Whale Culture: Whales and Beliefs), http://www.geocities.co.jp/NatureLand-Sky/3011/kujira-to-sinkou.html, accessed November 23, 2015

Appendix 5

Table A.5 Selected folklore legends concerning whales

Location	*Shrine/ temple*	*Description*
Miyagi Prefecture		
Karakuwa	Osaki Shrine	Ōji family and white whale
Fukui Prefecture		
Oshima	Ōminato Shrine	God and whale
Mie Prefecture		
Ōsatsu		Eleven-faced Goddess of Mercy and whale
Shiroura	Jōrin Temple	Spirit of pregnant whale (1759)
Wakayama Prefecture		
Taiji		Mother North Pacific right whale accompanied by calf
Yamaguchi Prefecture		
Senzaki		Tonomura family and spirit of mother whale
Saga Prefecture		
Yobuko		Spirit of mother whale with calf on pilgrimage to Benten Island
Nagasaki Prefecture		
Uku Island		Yamada Monkurō and whale with calf on pilgrimage to Daihō Temple (1716)

(*continued*)

M. Itoh, *The Japanese Culture of Mourning Whales*,
https://doi.org/10.1007/978-981-10-6671-9

Table A.5 (continued)

Location	*Shrine/ temple*	*Description*
Ojika Island		Oda Denbee and whale on pilgrimage to Ise Grand Shrine
Ojika Island		Oda Denbee and blue whale on its way to Ryūgū Palace (the undersea palace of the Dragon God of the Sea)
Uonome, Nakadōri Island		Yogorō and pregnant whale
Fukue Island		Yamada Monkurō and whale with calf on pilgrimage to Daihō Temple (1716)

Sources: Yoshihara Tomokichi, "Kujira no haka" (Graves of Whales), in Tanigawa Ken'ichi, ed., *Nihon minzoku bunka shiryō shūsei* (Collection of Documents on Japanese Folk Culture), Vol. 18, Tokyo: San'ichi-shobō, 1997, 468, 472–475; "Nihon no kujira bunka: Kujira ni kakawaru densetsu · itsuwa" (Japanese Whale Culture: Folklore Legends and Anecdotes Regarding Whales), http://www.geocities.co.jp/NatureLand-Sky/3011/33kujira-densetu.html, accessed December 23, 2015

Selected Bibliography

I Sources in English

Books

Ambros, Barbara R. 2012. *Bones of Contention: Animals and Religion in Contemporary Japan*. Honolulu: University of Hawai'i Press.

Ash, Christopher. 1962. *Whaler's Eye*. New York: Macmillan.

Asquith, Pamela J., and Arne Kalland, eds. 1997. *Japanese Images of Nature: Cultural Perspectives*. Richmond/Surrey: Curzon Press.

Boreal Institute for Northern Studies, ed. 1988. *Small-Type Coastal Whaling in Japan: Report of an International Workshop*. Edmonton: Boreal Institute for Northern Studies, the University of Alberta.

Burnett, D. Graham. 2007. *Trying Leviathan: The Nineteenth-Century New York Court Case That Put the Whale on Trial and Challenged the Order of Nature*. Princeton: Princeton University Press.

———. 2012. *The Sounding of the Whale: Science and Cetaceans in the Twentieth Century*. Chicago: University of Chicago Press.

Collcutt, Martin. 1981. *Five Mountains: The Rinzai Zen Monastic Institution in Medieval Japan*. Cambridge, MA: Council on East Asian Studies, Harvard University.

Dolin, Eric Jay. 2008. *Leviathan: The History of Whaling in America*. New York: W. W. Norton.

Dorsey, Kurkpatrick. 2014. *Whales and Nations: Environmental Diplomacy on the High Seas*. Seattle: University of Washington Press.

M. Itoh, *The Japanese Culture of Mourning Whales*,
https://doi.org/10.1007/978-981-10-6671-9

Elliot, Gerald. 1997. *Whaling 1937–1967: The International Control of Whale Stocks*, Kendall Whaling Museum Monograph Series, No. 10. Sharon: Kendall Whaling Museum.

Ellis, Richard. 1991. *Men and Whales*. New York: Knopf.

Garon, Sheldon M. 1987. *The State and Labor in Modern Japan*. Berkeley: University of California Press.

Hardacre, Helen. 1984. *Lay Buddhism in Contemporary Japan: Reiyukai Kyodan*. Princeton: Princeton University Press.

———. 1989. *Shinto and the State, 1868–1988*. Princeton: Princeton University Press.

Hrebenar, Ronald J., and Nakamura Akira, eds. 2015. *Party Politics in Japan: Political Chaos and Stalemate in the Twenty-first Century*. London: Routledge.

Kalland, Arne. 1995. *Fishing Villages in Tokugawa Japan*. Honolulu: University of Hawai'i Press.

———. 2009. *Unveiling the Whale: Discourse on Whales and Whaling*. New York: Berghan Books.

Kalland, Arne, and Brian Moeran. 1992. *Japanese Whaling: End of an Era*. London: Curzon Press.

Kishigami, Nobuhiro, and James M. Savelle, eds. 2005. *Indigenous Use and Management of Marine Resources*. Osaka: National Museum of Ethnology.

Kishigami, Nobuhiro, Hamaguchi Hisashi, and James M. Savelle, eds. 2013. *Anthropological Studies of Whaling*. Osaka: National Museum of Ethnology.

Knight, John, ed. 2004. *Wildlife in Asia: Cultural Perspectives*. London: RoutledgeCurzon.

Linzey, Andrew. 2009. *Why Animal Suffering Matters: Philosophy, Theology, and Practical Ethics*. New York: Oxford University Press.

———, ed. 2013. *The Global Guide to Animal Protection*. Champaign: University of Illinois Press.

Maxey, Trent E. 2014. *The "Greatest Problem": Religion and State Formation in Meiji Japan*. Cambridge, MA: Harvard University Asian Center.

Melville, Herman. 1966. *Moby Dick or, The Whale*. New York: Harper & Row. (Originally Published in 1851).

Miyaoka, Isao. 2004. *Legitimacy in International Society: Japan's Reaction to Global Wildlife Preservation*. New York: Palgrave Macmillan.

Morikawa, Jun. 2009. *Whaling in Japan: Power, Politics, and Diplomacy*. New York: Columbia University Press.

Nakamura, Hajime. 1964. *Ways of Thinking of Eastern Peoples: India, China, Tibet, Japan*. Trans. P.P. Wiener. Honolulu: East-West Center Press.

Pflugfelder, George M., and Brett L. Water, eds. 2005. *JAPANimals: History and Culture in Japan's Animal Life*. Ann Arbor: Center for Japanese Studies, University of Michigan.

Reader, Ian, Esben Andreasen, and Finn Stefansson, eds. 1993. *Japanese Religions Past and Present*. Honolulu: University of Hawai'i Press.

Shindō, Naosaku. 1975. *History of Whales in the Inland Sea*. Kobe: Hyōgo Dietists' Institute.

Stoett, Peter J. 1997. *The International Politics of Whaling*. Vancouver: UBC Press.

Taiji Whale Museum, ed. 2011. *Whale Boats: Styles and Designs*. Taiji: Taiji Whale Museum.

Tonnessen, J.N., and Johnsen, A.O. 1982. *The History of Modern Whaling*. Trans. R.I. Christophersen. Berkeley: University of California Press.

Young, Oran. 2010. *Institutional Dynamics: Emergent Patterns in International Environmental Governance*. Cambridge, MA: MIT Press.

Book Chapters

Abel, Jessamyn R. 2005. The Ambivalence of Whaling: Conflicting Cultures in Identity Formation. In *JAPANimals: History and Culture in Japan's Animal Life*, ed. George M. Pflugfelder and Brett L. Water, 315–340. Ann Arbor: Center for Japanese Studies, University of Michigan.

Kalland, Arne. 2004. Japanese Perceptions of Whales and Dolphins. In *Wildlife in Asia: Cultural Perspectives*, ed. John Knight, 73–87. London: RoutledgeCurzon.

Peterson, M.J. 1997. Whales, Cetologists, Environmentalists, and the International Management of Whaling. In *Knowledge, Power and International Policy Coordination*, ed. Peter M. Haas, 147–186. Columbia: University of South Carolina Press.

Journal Articles and Other Sources

Calder, Kent E. 1988. Japanese Foreign Economic Policy Formation: Explaining the Reactive State. *World Politics* 40 (4): 517–541.

Iwasaki-Goodman, Masami, and Masahiro Nomoto. 1999. The Ainu on Whales and Whaling. In *Spirit of a Northern People*, ed. W.W. Fitzhugh, 222–226. Washington, DC: National Museum of Natural History Smithsonian Institution.

Kalland, Arne. 2002. Holism and Sustainability: Lessons from Japan. *Worldviews: Global Religions, Culture, and Ecology* 6 (2): 145–158.

Kato, Kumi. 2007. Prayers for the Whales: Spirituality and Ethics of a Former Whaling Community—Intangible Cultural Heritage for Sustainability. *International Journal of Cultural Property* 14: 283–313.

Kato, Hidehiro, Motoi Yoshioka, and Seiji Ohsumi. 2005. Current Status of Cetaceans and Other Marine Mammals in the North Pacific, with a Review of Advanced Research Activities on Cetacean Biology in Japan. *Mammal Study* 30: 113–124.

McCarney, Perry. 2010. Japan's Standpoint on Whaling. http://www.helium.com/items/1516632-japan-view-on-whaling

Miyaoka, Isao. 2016. *State-Society Relations Under Foreign Pressure: Two Contrasting Cases in the Japanese Fisheries Policy*, 135–161. http://www.dijtokyo.org/doc/dij-jb_11-miyaoka.pdf. Downloaded February 24, 2016.

II Sources in Japanese

Books

Akimichi, Tomoya. 1994. *Kujira to hito no minzoku-shi* (Ethnography of Whales and People). Tokyo: Tokyo daigaku-shuppankai.

———. 2009. *Kujira wa dareno monoka* (Who Owns Whales). Tokyo: Chikuma-shinsho.

———. 2013. *Umi ni ikiru kaijin no minzokugaku* (Ethnology of People Who Live in the Sea). Tokyo: Tokyo daigaku-shuppankai.

Fukumoto, Kazuo. 1960. *Nihon hogeishi-wa* (History of Japanese Whaling). Tokyo: Hōsei daigaku-shuppankai.

Hirado-shi kyōiku-iinkai, ed. 1980. *Saikai geigei-ki* (Record of Whaling in the West Sea). Hirado: Hirado-shi kyōiku-iinkai.

Kaneko, Misuzu. 1984. *Kaneko Misuzu zenshū* (Complete Works of Kaneko Misuzu), vol. 3. Tokyo: JULA shuppan-kyoku.

Kawano (not Kōno), Ryōsuke. 2005. *Chōshū · Kitaura hogei no aramashi* (Introduction to Chōshū Northern Bay Classical Whaling). Nagato: Nagato-Ōtsu kujira-shokubunka o keishō suru-kai.

Kishigami, Nobuhiro. 2003. *Kaiyō-shigen no riyō to kanri ni kansuru jinruigaku-teki kenkyū* (Anthropological Study of Use and Management of Marine Resources). Osaka: Kokuritsu minzoku hakubutsukan.

———. 2012. *Hogei no bunka-jinruigaku* (Anthropology on Whaling). Tokyo: Seizandō-shoten.

Komatsu, Masayuki. 2005. *Kujira sono rekishi to bunka* (History and Culture of Whales). Tokyo: Goma-shobō.

———. 2007a. *Nihonjin to kujira* (The Japanese and Whales). Tokyo: Goma-shobō.

———. 2007b. *Yutakana Tokyo-wan: Yomigaere Edo-mae no umi to shoku-bunka* (Rich Tokyo Bay: Restore the Sea and Food Culture of Edo). Tokyo: Yūzankaku.

———. 2010. *Sekai kujira sensō* (World War on Whales). Tokyo: PHP kenkyūjo.

Kon'no, Tsutomu. 2007. *Kaneko Misuzu futatabi* (Revising Kaneko Misuzu). Tokyo: Shōgakukan, pbk., 2011.

Kumano Taiji-ura hogei-shi hensan-iinkai, ed. 1965. *Kumano no Taiji: Kujira ni idomu machi* (Taiji, Kumano: Town That Challenges Whales). Tokyo: Heihonsha.

Morita, Katsuaki. 1994. *Kujira to hogei no bunkashi* (Cultural History of Whales and Whaling). Nagoya: Nagoya daigaku-shuppankai.

Nakamura, Ikuo. 1988. *Kami to hito no seishin-shi: Nihon Bukkyō no shinsō-kōzō* (Spiritual History of Gods and People: Deep Core Structure of Japanese Buddhism). Kyoto: Jinbun-shoin.

Nakamura, Hajime. 1995. *Daijō Bukkyō no shisō* (Philosophy of Mahayana Buddhism). Tokyo: Shunjūsha.

Nakamura, Ikuo. 2001. *Saishi to kugi: Nihonjin no shizenkan · dōbutsukan* (Religious Dedications and Sacrifices: Japanese Views of Nature and Animals). Kyoto: Hōzōkan.

Nakamura, Hajime. 2010a. *Jihi* (Benevolence). Tokyo: Kōdansha.

Nakamura, Ikuo. 2010b. *Nihonjin no shūkyō to dōbutsu-kan: Sessho to nikujiki* (Religion and Views of Animals of the Japanese: Killing and Meat-Eating). Tokyo: Yoshikawa-Kōbunkan.

Nakazono, Shigeo. 2000. *Ikitsuki-jima no kakure kirishitan* (Hidden Christians on Ikitsuki Island). Nagasaki: Ikitsuki-chō hakubutsukan · Shimano yakata.

———. 2001. *Kujira-tori no keifu: Gaisetsu Nihon hogei-shi* (Genealogy of Whaling: Introduction to the History of Japanese Whaling). Nagasaki: Nagasaki Shimbunsha.

Nakazono, Shigeo, and Yasunaga, Hiroshi. 2009. *Kujira-tori e-monogatari* (Picture Tales of Whaling). Fukuoka: Gen-shobō.

Natori, Takemitsu. 1945. *Funka-wan Ainu no hogei* (Ainu Whaling in Funka Bay). Sapporo: Hoppō-bunkasha.

Nihon geirui kenkyūjo and Nihon hogei kyōkai, eds. 2003. *Dai-1-kai Nihon dentō-hogei chiiki samitto kaisai no kiroku* (Conference Proceedings of First Regional Summit on Japanese Traditional Whaling). Nagato: Nagato-shi.

———. 2005a. *Dai-2-kai Nihon dentō-hogei chiiki samitto kaisai no kiroku* (Conference Proceedings of Second Regional Summit on Japanese Traditional Whaling). Hirado: Ikitsuki-chō.

———. 2005b. *Dai-3-kai Nihon dentō-hogei chiiki samitto kaisai no kiroku* (Conference Proceedings of Regional Summit on Japanese Traditional Whaling). Muroto: Muroto-shi.

———. 2006. *Nihon dentō-hogei chiiki samitto: Kaisai no kiroku* (Regional Summit on Japanese Traditional Whaling: Conference Proceedings). Shimonoseki: Shimonoseki-shi.

———. 2007. *Nihon dentō-hogei chiiki samitto: Kaisai no kiroku* (Regional Summit on Japanese Traditional Whaling: Conference Proceedings). Wakayama: Taiji-chō.

Ohsumi, Seiji. 1988. *Kujira wa mukashi riku o aruite-ita: Shijō saidai no dōbutsu no shinpi* (Whales Used to Walk on the Land: The Biggest Mystery about Animals in History). Tokyo: PHP kenkyūjo.

———. 2003. *Kujira to Nihonjin* (Whales and the Japanese). Tokyo: Iwanami-shinsho.

———. 2008. *Kujira o otte han-seiki: Shin-hogei-jidai heno teigen* (Half-Century of Studying Whales: Proposals for New Age of Whaling). Tokyo: Seizandō-shoten.

Ohsumi, Seiji, and Katō, Hidehiro, eds. 2002. *Geirui shigen no jizokuteki shiyō ha kanō ka: Geirui shigen kenkyū no saizensen* (Is Sustainable Use of Cetacean Resources Possible?: The Frontline of Cetacean Resources Study). Tokyo: Seibutsu-kenkūsha.

Shindō, Naosaku. 1968. *Seto-naikai no kujira no kenkyū* (Study of Whales in the Inland Sea). Kobe: Kobe-shi ishi-kyōdo-kumiai.

———. 1970. *Seto-naikai shūhen no kujira no kenkyū* (Study of Whales Surrounding the Inland Sea). Kobe: Ikuta-ku ishikai.

Taiji-chōritsu kujira no hakubutsukan, ed. 2011. *Kujira-bune: Katachi to ishō* (Whale Boats: Styles and Designs). Taiji: Taiji Whale Museum.

Takahashi, Jun'ichi. 1992. *Kujira no Nihon bunka-shi: Hogei-bunka no kōseki o tadoru* (Japanese Cultural History of Whales: Tracing the Paths of Whaling Culture). Kyoto: Tankōsha.

Tanigawa, Ken'ichi. 1975. *Kami · Ningen · Dōbutsu: Denshō o ikiru sekai* (Gods · People · Animals: World of Living Folklore). Tokyo: Heibonsha.

———, ed. 1997. *Nihon minzoku bunka shiryō shūsei* (Collection of Documents on Folk Culture in Japan), vol. 18. Tokyo: Sanichi-shobō.

Ueda, Yoshie. 1979. *Nagato nisō-monogatari* (Tale of Nuns in Nagato). Tokyo: Kokusho-kankōkai.

Wang, Min. 2006. *Nitchū 2000-nen no furikai: Kotonaru bunka "kisō" o saguru* (Two Thousands Years of Misunderstanding Between China and Japan: To Grope for "Fundamental Stratum" of Cultural Difference). Tokyo: Asahi Shinsho.

Watanabe, Hiroyuki. 2006. *Hogei-mondai no rekishi shakaigaku: Kingendai Nihon ni okeru kujira to ningen* (Historical Sociology of Whaling Problems: Whales and People in Modern Japan). Tokyo: Tōshindō.

Yamakawa, Tōru. 2010. *Toruka mamoruka?: Kujira no mondai* (To Catch or Protect?: Problems of Whaling). Tokyo: Gijutsu-hyōronsha.

Yazaki, Setsuo. 1993. *Dōyō-shijin Kaneko Misuzu no shōgai* (Life of the Poet Kaneko Misuzu). Tokyo: JULA shuppan-kyoku.

Yoshihara, Tomokichi. 1982. *Bōnan hogei: Bōsō-chihō no hogei shiryō* (Whaling in South Bōsō: Documents on Whaling in Bōsō Region), Tokyo: Aizawa-bunko.

Book Chapters

Yoshihara, Tomokichi. 1997. Kujira no haka (Graves of Whales). In *Nihon minzoku bunka shiryō shūsei* (Collection of Documents on Japanese Folk Culture), ed. Tanigawa Ken'ichi, vol. 18, 409–478. Tokyo: Sanichi-shobō.

Journal Articles and Online Publications

Fujii, Fuminori. 2013. 'Nagato gei-rei sūhai densetsu' o tadasu (To Question the "Legend about the Whale Spirit in Nagato"). *Kyōdo Bunka Nagato* 25: 34–64.

———. 2014. Ōhibi Amadera Hōsen-an no rekishi: Jikō-Shōshin anshu no inori (History of Nunnery Hōsen-an in Ōhibi: Prayer of Abbess Jikō-Shōshin). *Kyōdo Bunka Nagato* 26: 70–110.

Harada, Shin'ichi. 1994. Hirugo [*sic*] shinwa ronkō (Analysis of Hirugo Mythology), 147–194. http://repo.komazawa-u.ac.jp/opac/repository/all/16167/KJ00005086929.pdf

Iwamoto, Chigusa. 2008. Jōdo-shū Amadera: Hōsen-an no seikatsu to shokuji (Jōdo School Nunnery: Life and Diet of Hōsen-an). *Kyōdo Bunka Nagato* 20: 78–91.

Iwasaki, Masami. 2002. Ainu-minzoku kujira-riyō-bunka no sokuseki o tadoru (Tracing Whale Utilization Culture of Ainu People). *Hokugakuen-daigaku jinbun-ronshū* 21: 111–146.

Kon'no, Tsutomu. 2012. Kaneko Misuzu ga nokoshita-mono (Legacy of Kaneko Mizusu). *Kyōdo Bunka Nagato* 24: 132–149.

"Kujira: Nihon no koshiki-hogei to zusetsu." 2011. *Whales: Japanese Classical Whaling and Illustrations.* http://www.lib.u-tokyo.ac.jp/tenjikai/tenjikai2011/kujira.html

Ohsumi, Seiji. 2005. Geirui to gyogyō no kyōgō mondai (Problems of Competition Between Cetaceans and Fisheries). *Shoku no kagaku* 327: 34–40.

———. 2013a. Kokusai hogei-iinkai/Kagaku shō-iinkai no hensen to Nihon tono kankei (I) Senzen no kokusai hogei kisei to kagaku no kanyo (History of International Whaling Commission Science Subcommittee and Its Relations with Japan, Part I: International Regulations for Whaling During Pre-World War II [Era] and Its Involvement of Science). *Geiken-tsūshin* 458: 1–7.

———. 2013b. Kokusai hogei-iinkai/Kagaku shō-iinkai no hensen to Nihon tono kankei (II) Kokusai hogei torishimari jōyaku no setsuritsu to Nihon no kamei (History of International Whaling Commission Science Subcommittee and Its Relations with Japan, Part II: Establishment of International Convention for the Regulation of Whaling and Japan's Admission). *Geiken-tsūshin* 459: 1–9.

Tsunoda, Fusako. 1969. Ōmi-jima: Nisō no iru chi no hate (Ōmi Island: The Edge of the World Where Nuns Live). *Tabi*: 70–74.

Yoshihara, Tomokichi. 1976. Bōnan bogei (Whaling in South Bōsō). *Tokyo suisan-daigaku ronshū* 3 (Bessatsu): 17–144.

Index

M. Itoh, *The Japanese Culture of Mourning Whales*,
https://doi.org/10.1007/978-981-10-6671-9

K

O

P

FSC
www.fsc.org
MIX
Papier | Fördert gute Waldnutzung
FSC® C083411

Kolibri 360 GmbH
Ferdinand-Jühlke-Straße 7
99095 Erfurt, Deutschland
produktsicherheit@kolibri360.de